Discovery EDUCATION | MATH TECHBOOK

GEOMETRY WITH HONORS

TEACHER GUIDE FOR FLORIDA INTERACTIVE STUDENT EDITION

Log in to Math Techbook at
www.DiscoveryEducation.com

ISBN 13: 978-1-68220-487-0

Library of Congress Cataloging-in-Publication Data available upon request from the Library of Congress.

Printed in the United States of America.

1 2 3 4 5 6 7 8 9 10

Discovery EDUCATION™

800-323-9084
One Discovery Place, Silver Spring, Maryland 20910

Table of Contents

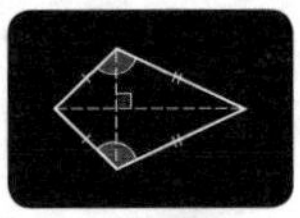

UNIT 9: Quadrilaterals and Other Polygons

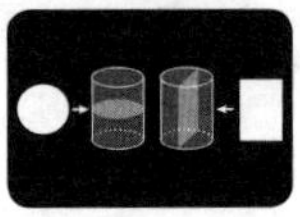

UNIT 10: 3-D Figures

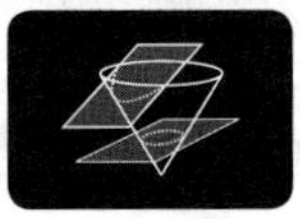

UNIT 11: Conic Sections

Letter to the Teacher

Dear Teacher,

Welcome to Math Techbook™! This Teacher Guide provides an overview of Discovery Education Math Techbook and contains answers to questions that appear in the *Interactive Student Edition*. This guide will help you explore the instructional techniques and conceptual framework of Math Techbook, as well as prepare you for your key role as a facilitator of mathematics learning.

In this Teacher Guide, you'll find two major sections:

- OVERVIEW: Math Techbook is different from other programs you've used, and this section will help you understand how and why. In addition, you'll learn how to navigate Math Techbook and effectively use the Discover-Practice-Apply instructional cycle.
- ANSWER KEYS: The *Interactive Student Edition* contains more than 500 questions, and answers for each question appear in this Teacher Guide. Links are also provided to solutions for the Apply problems, which are typically more complex multistep problems.

The *Interactive Student Edition* is not the primary instructional resource. Rather, it is a print subset of Math Techbook, a fully digital, core instructional resource. Math Techbook engages students with dynamic content designed to help students develop deep mathematical understanding and apply it to the world around them. As the classroom facilitator, you'll be equipped with a comprehensive resource that includes student activities, model lessons, and a dashboard that shows student progress—all in one, easy-to-navigate site.

Math Techbook provides the all-important learning opportunities that encourage students to develop conceptual understanding, procedural fluency, and application in a structured and balanced manner. It offers many digital resources, none of which would be available in a print curriculum or PDF-driven sites:

- MATH TOOLS: A suite of digital tools, including a graphing calculator, white board, and dynamic geometry software, are included with Math Techbook.
- INTERACTIVES: Within Math Techbook, students are exposed to online activities that allow the user to interact with them, playing games that involve everything from skeleton armies and absolute value in Grade 6 to caffeine elimination and exponential decay in Algebra II.
- VIDEOS: Because you expect nothing less from Discovery Education, Math Techbook includes a vast assortment of videos that engage students in new topics, present a problem situation, or explain concepts.
- TECHNOLOGY-ENHANCED ITEMS (TEIs): These computer-delivered questions allow special interactions for collecting data, going far beyond the typical multiple-choice and constructed-response items of the past. All data gathered by the TEIs is reported in the dashboard.
- DASHBOARD: Students, teachers, and parents can monitor student progress using real-time data.

What Makes the ISE Interactive?

As previously mentioned, the *Interactive Student Edition* is not a complete instructional tool. Rather, it is designed for students to interact with the comprehensive digital Math Techbook. Quick-response (QR) codes are provided throughout the resource to direct students to specific learning resources.

To use the QR codes, you will need a QR reader. Readers are available for phones, tablets, laptops, desktops, and virtually any device in between. Download a free QR reader in the App Store or Google Play. To access Math Techbook resources, follow these steps:

1. Open the QR code reader on your device.
2. Hold your device so the QR code is visible within your device's screen. One of two things will happen:
   - The device may automatically scan the code; or,
   - The device will scan the code when you press a button, similar to taking a picture.

3. Once scanned, the QR code will direct you to a page or resource on the web.
4. For resources in Math Techbook, you may need to log in with your username and password the first time you access a QR code. After that, you won't need to log in again, unless you sign out or remain inactive for too long.

Scan this QR code to access a video that provides a deeper introduction to Math Techbook:

As you implement Math Techbook and its helpful digital content, prepare for a transformation! Enjoy the journey, and best of luck for a fantastic year!

Sincerely,

The Discovery Education Math Team

How to Use Math Techbook™

The Techbook Difference

Discovery Education Math Techbook is a dynamic, digital resource that empowers students to take ownership of their learning. Compatible with any hardware used by school districts, Discovery Education Math Techbook elevates student achievement through a combination of three critical elements.

I. **Content:** Exclusive content from Discovery Education engages students and brings math to life, while meeting all the demands of the standards.

II. **Instruction:** Research-based, inquiry-driven instructional design balances conceptual understanding, procedural fluency, and application to relevant problems.

III. **Technology:** Powerful, simple-to-use technology saves teachers time and eases the transition to digital resources.

It is the convergence of all three elements that makes Math Techbook truly unique. Math Techbook is a powerful resource that both inspires today's digital native students and brings transformational change to the classroom. The result is a marked increase in student engagement and learning.

The Power of Three

Navigating Math Techbook

After signing in and navigating to your desired Math Techbook course, **select a unit** you would like to teach. The units in each grade level provide in-depth coverage of the standards.

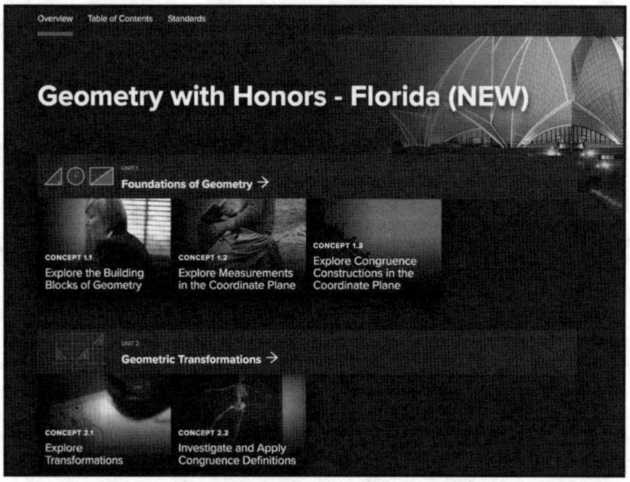

Within each unit are one or more concepts to explore. Each unit also includes a Unit Opener that presents a real-world situation related to the unit's focus. **Select a concept** by clicking on a concept title or an image.

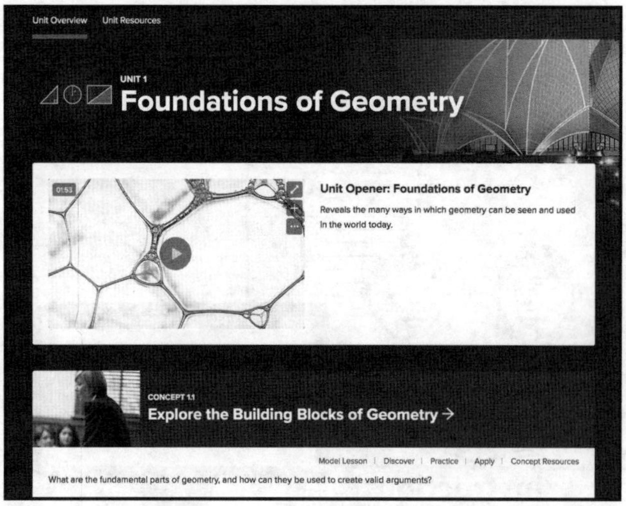

Navigate using the Discover, Practice, Apply, and Model Lesson tabs below the title to **explore the components** of a concept. The Model Lesson provides all of the information you need to successfully teach each concept: step-by-step suggestions for using the interactives, questions for discussion, and assessment items that will assist your students in the exploration and deep learning of a concept.

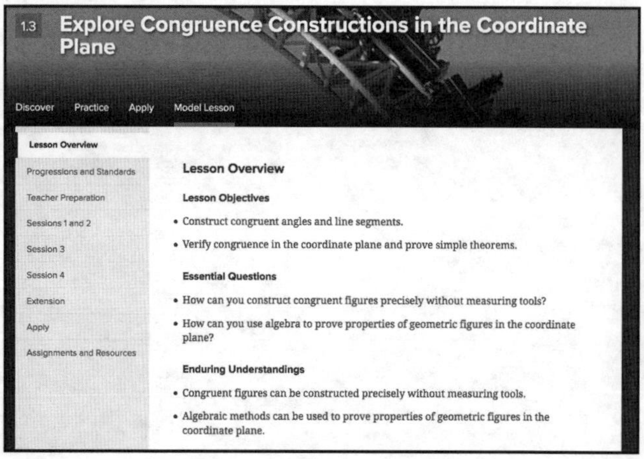

Within Discover, **explore the content** that your students will engage with. Through the activities in Engage and Investigate, students will be presented with real-world problems. They will use interactives and digital tools to make sense of problems and find solutions.

After completing the investigations, students will practice their new skills in a game-like environment and apply their skills in a way that requires more rigor and deeper connections. You can view exercises for procedural fluency in Practice and view real-world problems worth solving in Apply.

Discovery Education's Math Techbook also includes tools that assist students' learning and help you keep track of each student. The **Teacher Learning Dashboard** shows you student results. Students also have a version of the Dashboard where they can review their work.

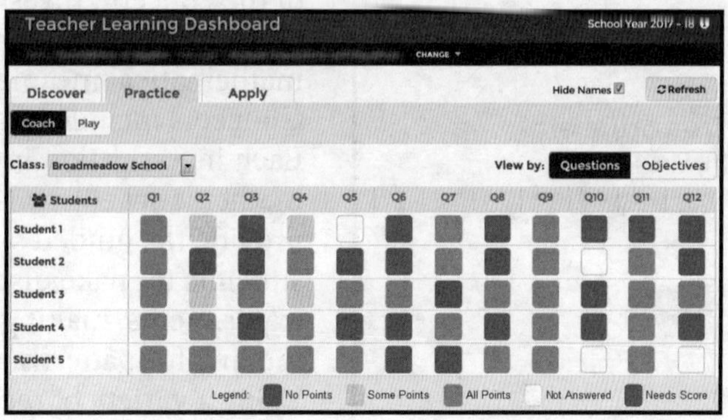

Math Tools include a graphing calculator, dynamic geometry software, and other tools your students can use to explore math interactively. Many of these tools are embedded within lessons, but you can always choose to use them in your own way and have students share their work with you or the entire class.

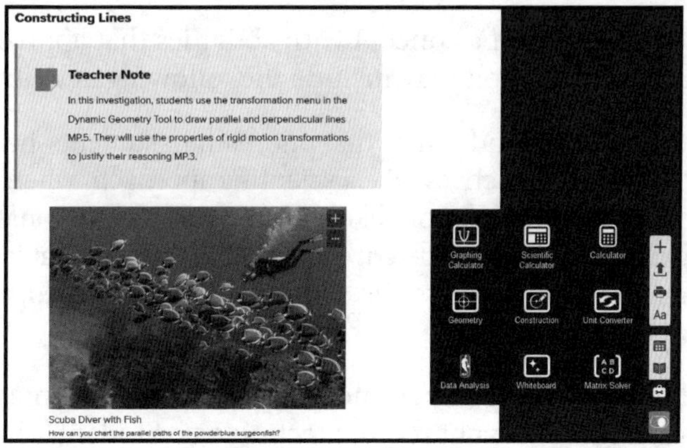

Discover – Practice – Apply

Discovery's Math Techbook is based on a Discover, Practice, Apply instructional cycle. This cycle of learning enables students to solve problems in more than one way using more than one modality, allowing students to reach mastery learning with more depth. Students retain more of what they have learned.

Discover

Engage: An intriguing video or thought-provoking image and situation captures students' attention and ignites their curiosity. This activity motivates students to learn while providing an opportunity to activate prior knowledge, connecting back to concepts and terms previously learned. The first page of each concept in the *Interactive Student Edition* introduces students to the skills and topics they will encounter. Each concept includes one or more Essential Questions that will help students focus on the key ideas. Students may return to the Essential Questions as a form of self-assessment at the end of each concept.

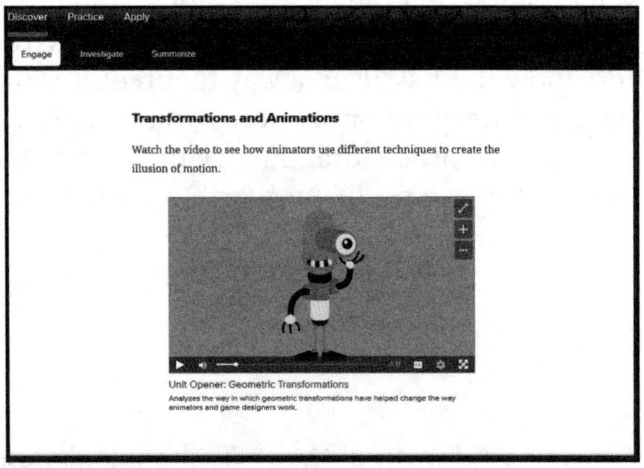

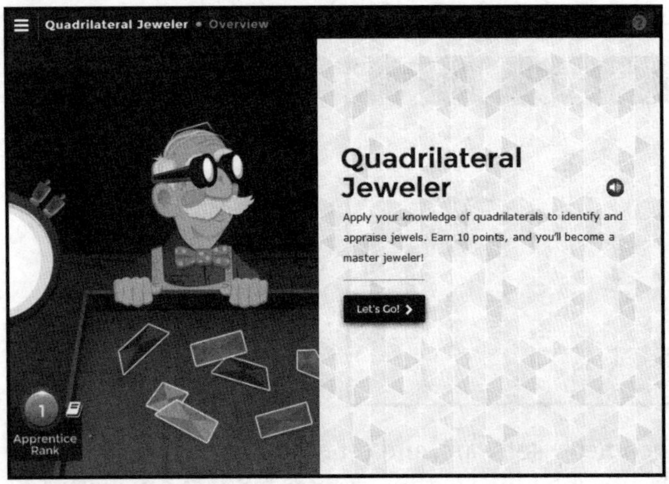

Investigate: This section is the main component of the concept. Investigations keep students actively engaged, thinking, and developing mathematical meaning in everyday settings. Concepts include two to six investigations. Each investigation immerses students in mathematical or real-world scenarios. They provide the guided support necessary for students to make discoveries by conducting experiments, making observations, testing ideas, looking for patterns, and summarizing results. Using a variety of recommended instructional and classroom management strategies, students deepen their understanding by sharing their discoveries, listening to the conjectures of their classmates, and critiquing arguments.

Investigations scaffold the development of a concept with activities that approach the concept in different ways. The activities within any investigation may include the following activity types:

- **Interactives** are exploratory activities and games that put math in students' hands and allow them to explore a concept in depth. Some interactives take a scientific approach, where students create several test cases and compare results. Others feel more like a game, with motivational elements that encourage strategic, critical thinking. Students learn by discovering math concepts that lead to successful outcomes. As a follow-up, students analyze patterns, describe connections they noticed, and summarize what they learned in a formal way.

- **Hands-On Activities** allow students to experience and discover mathematical relationships in ways similar to the interactives, but they do not require technology. Students can work independently or collaboratively in small groups.

- **Technology-Enhanced Items** encourage students to summarize their thinking and allow them to become familiar with new question formats they will see on high-stakes assessments.

In the *Interactive Student Edition*, students can record the results of their learning from within Engage and Investigate in their own words. Doing so may bring up further questions, giving you the opportunity to address any misconceptions and help students solidify their learning.

To give students practice and to confirm student understanding, the *Interactive Student Edition* includes a Check for Understanding for each investigation. Used as a nightly assignment, Check for Understanding assesses the depth of knowledge your students retained from the day's learning.

The *Interactive Student Edition* also includes a summary of key concepts and skills, a condensed version of what is available at the end of each concept. When students have a partial understanding of the concept and require reinforcement, the Summary section gives students the opportunity to confirm what they learned during the investigations and address misconceptions that may return when students are working independently.

Practice

The Practice section consists of two parts, Coach and Play, each containing a collection of exercises for your students to sharpen their skills.

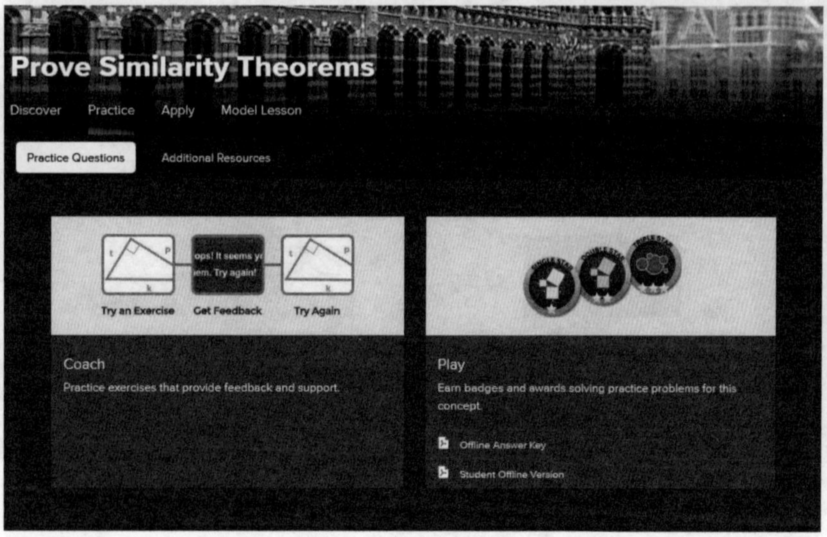

The items within **Coach** provide a set of guided practice exercises. Immediate, specific feedback in the form of tips or suggestions directly related to a student's error is provided when an incorrect answer is entered. Feedback often directly addresses the student's misconception and always guides him or her to the correct solution path. If a student is unable to identify the correct answer in three attempts, the correct answer is provided with instruction. This allows students to continue to learn as they answer each question. Similar to athletics, Coach provides necessary support, giving suggestions, advice, just-in-time instruction, and a virtual high-five for student success.

The items within **Play** provide a set of independent practice exercises. Just as in athletics, the coach cannot step in and provide assistance during the big game. The player's success is up to his or her own strength and abilities. In Play, students answer questions independently to earn badges for their knowledge. Students earn a star badge for the concept after successfully completing five problems, and they get more stars added to their badge as they correctly solve more problems. Students are motivated to continue solving problems to improve their score and earn more stars.

The *Interactive Student Edition* contains a set of Practice Exercises, and a printable version of the Play items is available online through the Student Offline Version button. Consequently, there are two options for students to practice their skills when they're away from the computer.

Apply

Apply culminates the learning cycle by ensuring that students not only understand the concept, but can also use what they've learned in their everyday lives. The problems in Apply go beyond traditional word problems. Just as in real-world situations, all the information that students need may not be provided. Students are often required to conduct research to complete the problem, determine what information they need, and ignore information they do not need. Apply provides real-world scenarios that involve multiple steps and reasoning, and students must submit models, equations, drawings, or explanations to support their solutions. Most importantly, Apply situations often have more than one feasible solution, depending on the model that students use and the data they collect. Math Techbook provides teacher support for every Apply scenario by including a rubric with 4-click scoring of student solutions submitted online. Apply can be used as homework, formative or summative assessment, or for project-based assessment.

One Apply problem is included in the *Interactive Student Edition*, making it easy to use for offline assignments.

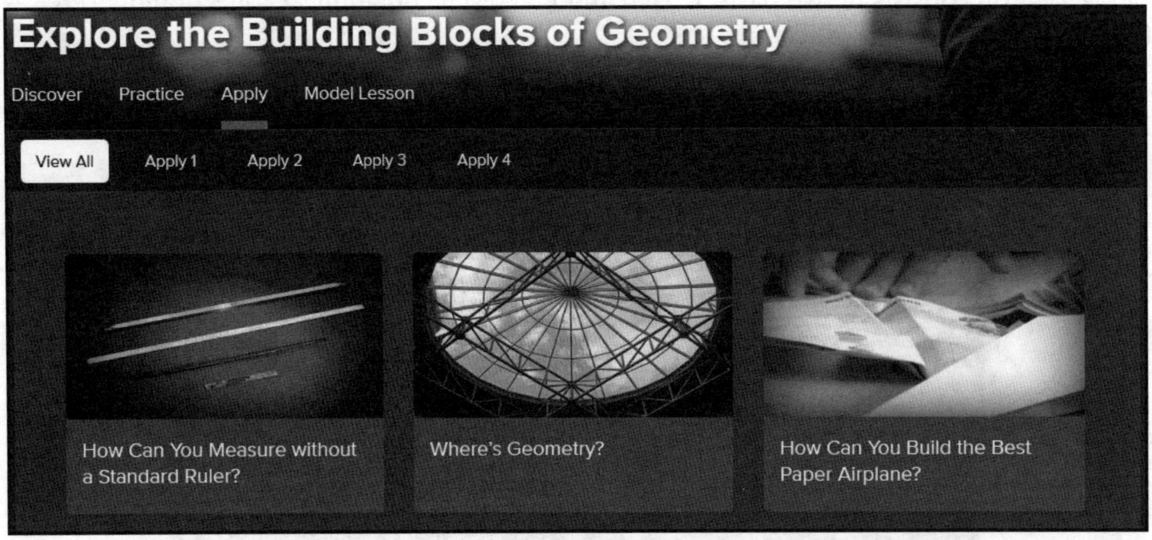

Discovery EDUCATION | MATH TECHBOOK

Answer Keys for Check for Understanding and Practice Exercises

1.1 Explore the Building Blocks of Geometry

Check for Understanding: Investigation 1

1.

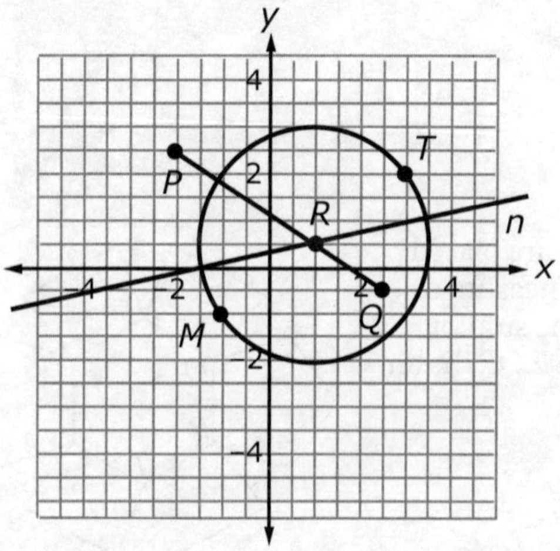

2. A, D, E
3. A. False
   B. True
   C. True
   D. True
4. Precision allows you to clearly communicate a mathematical argument to demonstrate the truth of a statement beyond any doubt.

Check for Understanding: Investigation 2

1. P = 16
2. $6 + 2n$ or $2(3 + n)$
3. A. Sample response: 33
   B. Sample response: The vertices of a triangle are all in the same plane, but they are not collinear.
4. A postulate is a statement that is accepted as true without proof. A conjecture is a statement based on observations that may or may not be true. A theorem is a statement that has been proved true.

Explore the Building Blocks of Geometry *(continued)*

Check for Understanding: Investigation 3

1.

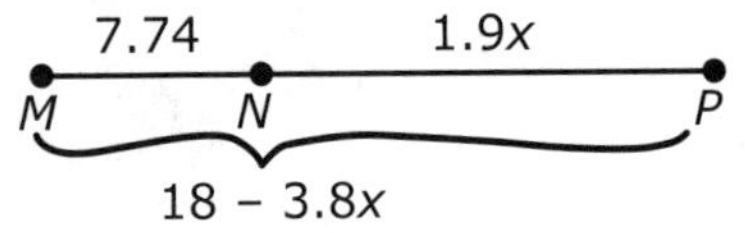

2. $x = 1.8; MP = 11.16$

3. Answers will vary. The sides and vertices of Miguela's polygons are placed so that they don't overlap. By the angle addition postulate, the measure of the new angle formed by the combination of two polygons is the sum of the angles. For example. $\angle PRQ$ is an angle of the pentagon and $\angle QRT$ is an angle of the hexagon. So $m\angle PRT = m\angle PRQ + m\angle QRT$.

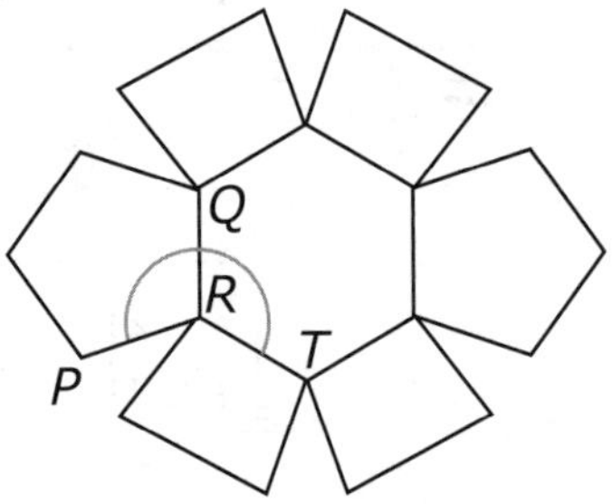

Check for Understanding: Investigation 4

1. B, A, C, B, C

2. Sample response: $AM = MD$. (*Note*: students will likely organize their thoughts in paragraph form rather than 2-column.)

Statement	Reason
I know that M is the midpoint of $\overline{AC}$ and $\overline{BD}$ and that $AC = BD$.	both are given
I can rewrite the lengths of the segments as $AC = AM + MC$ and $BD = BM + MD$.	segment addition postulate
So I can write $AC = BD$ as $AM + MC = BM + MD$.	substitution
Because M is the midpoint of $\overline{AC}$ and $\overline{BD}$, $AM = MC$, and $BM = MD$.	definition of midpoint
Now I can write $AM + MC = BM + MD$ as $2AM = 2MD$.	substitution
I can conclude that $AM = MD$.	division property

Practice Exercises

1. A, C, D
2. 31
3. D
4. D
5. C
6. 3
7. A, C
8. 43
9. B
10. $x = 10$
    $AB = 100$
    $MB = 50$
11. C, A, B

Apply

Where's Geometry?

This problem allows students to demonstrate their ability to articulate common geometric postulates and identify common geometric terms.

- If students have trouble understanding what it means for the geometric objects to exist in the physical world, then direct them back to the definition of the terms or the meanings of the postulates. There are subtleties that need to be considered when trying to take theoretical objects such as point, line, and plane and understand that they do not exist in the physical world.
- If another photo is not available, students can use the introductory image for the activity.

The Apply problems are rich performance tasks that may have a variety of open-ended solutions. One sample student response for this problem can be found in the Model Lesson.

1.2 Explore Measurements in the Coordinate Plane

Check for Understanding: Investigation 1

1. 16.44 units, 10.5 units^2
2. C, D
3. Student 1 divided polygon *ABCDE* into rectangles and triangles and added their areas. Student 2 drew a square around the polygon and found the area of the square, then subtracted the areas of the triangles and rectangle inside the square but outside the polygon. I prefer Student 2's method, because it's easier to draw. It's more difficult to figure out how to divide the inside of the polygon.

Check for Understanding: Investigation 2

1. A, B, E
2. $T(3.2, -4)$
3. *A* is the midpoint of $\overline{MN}$, so the coordinates of *A* are $\left(\frac{-1+-1}{2}, \frac{-1+3}{2}\right)$. *B* is the midpoint of $\overline{MR}$, so the coordinates of *B* are $\left(\frac{-1+4}{2}, \frac{-1+-1}{2}\right)$. Since *MACB* is a rectangle, *C* is the point of intersection of the vertical line through *B*, which has the equation $x = \frac{-1+4}{2}$ and the horizontal line through *A*, which has the equation $y = \frac{-1+3}{2}$. So, the coordinates of *C* are $\left(\frac{-1+4}{2}, \frac{-1+3}{2}\right)$,which is the midpoint of $\overline{MP}$.

Check for Understanding: Investigation 3

1. $F(1,4)$
2. B
3. E
4. Both ratios mean that you divide $\overline{MN}$ into 9 parts of the same length. The ratio 2 : 7 means you place a point on $\overline{MN}$ a distance equal to 2 parts from *M*, but the ratio 7 : 2 means you place a point on $\overline{MN}$ a distance equal to 7 parts from *M*. For example, suppose the endpoints of $\overline{MN}$ are $M(0, 0)$ and $N(9, 9)$. The point (2, 2) would partition $\overline{MN}$ into a ratio of 2 : 7, while (7, 7) would partition $\overline{MN}$ into a ratio of 7 : 2.

Practice Exercises

1. A
2. A
3. 5, 3, 2
4. B, C
5. C
6. A, B
7. B, E
8. 338
9. B, F
10. C

Explore Measurements in the Coordinate Plane *(continued)*

Apply

Why Is This Shape Special?

This problem allows students to apply what they know about ratios and geometric figures to identify a relationship among the lengths of the sides and segments of a golden rectangle.

- If possible, have students work on the activity independently so they can discover the ratio for themselves.
- If students will not be researching the ratio they find, provide this background information: This ratio is known as the Golden Ratio. Because the ratio of the length to the width of the rectangle is in this ratio, this is called a Golden Rectangle. This irrational number is an important constant, so it has been assigned its own name, the Greek letter Phi (φ). The Golden Ratio has some very interesting mathematical properties.
- Students can upload a snapshot of the image into the Dynamic Geometry Tool to superimpose the grid and determine the ratios, or they can copy or print the rectangle and complete the activity by hand.

The Apply problems are rich performance tasks that may have a variety of open-ended solutions. One sample student response for this problem can be found in the Model Lesson.

1.3 Explore Congruence Constructions in the Coordinate Plane

Check for Understanding: Investigation 1

1. D
2. B, E, F
3. To copy $\triangle ABC$, first use a compass and straightedge to construct $\angle A$. Next use a compass to mark off the length of $\overline{AB}$ and $\overline{AC}$, and construct these sides on the new triangle. Then use a compass and straightedge to construct $\angle C$. Finally, use the compass to double check that the length of $\overline{BC}$ matches the length of the third side in the new triangle.

Check for Understanding: Investigation 2

1. 5
2. B, C, E
3. Michelle uses the angle addition postulate twice. Because $\overleftrightarrow{JF}$ is perpendicular to the wall, $m\angle JFH + (5w - 13) = 90$. Then because the angle of incidence is equal to the angle of reflection, she uses the given information $m\angle GFH = 95°$ and $m\angle GFH = m\angle GFJ + m\angle JFH$ to conclude that $m\angle JFH = 0.5(95)$. In William's solution, he uses the fact that $\overleftrightarrow{AB}$ is a straight line and the angle addition postulate to reason that $180 = m\angle AFG + m\angle GFH + m\angle HFB$. Then because $m\angle GFH = 95°$, he concludes $180 - 95 = m\angle AFG + m\angle HFB$. William then uses the congruent complements theorem to write $m\angle AFG + m\angle HFB = 2(5w - 13)$.

Check for Understanding: Investigation 3

1. True, True, False
2. A, D
3. First, I would plot the center and point P in a coordinate plane. Next, I would draw a horizontal and a vertical segment to form a right triangle with the center and point P. Using the Pythagorean theorem, I would then calculate the length of the segment with endpoints at the center of the circle and P. If the length of the segment is more than $\sqrt{13}$, then this will prove that P must lie outside of the circle.

Explore Congruence Constructions in the Coordinate Plane *(continued)*

Practice Exercises

1. $-\frac{3}{4}, \frac{4}{3}$, yes, yes
2. 3, 3, yes
3. 13.25, 46
4. $-\frac{1}{3}$, 3, yes, yes
5. D
6. 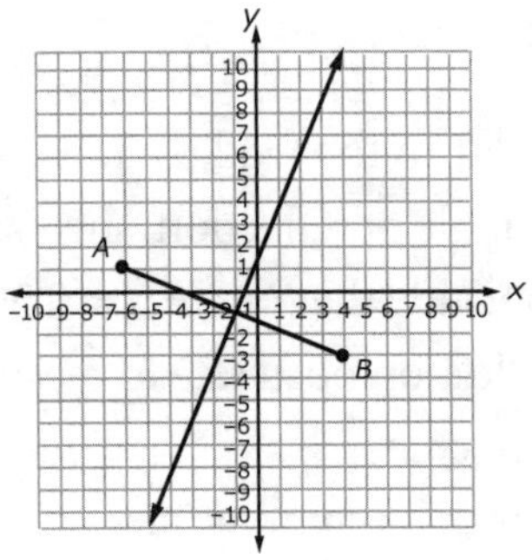

7. B, C
8. C

Apply

Is the Logo Ready for Production?

This problem challenges students to draw a right isosceles triangle and then use coordinate geometry including the distance formula and slope formula to prove its properties. Students then use geometric constructions to copy the triangle.

- Students may use the Dynamic Geometry Tool to create their logo. This activity can also be completed offline with graph paper, pencil, compass, and straightedge.
- If you use this problem as a homework assignment, you may want to set up the problem during class time by having students watch the video Jersey Logo.

The Apply problems are rich performance tasks that may have a variety of open-ended solutions. One sample student response for this problem can be found in the Model Lesson.

2.1 Explore Transformations

Check for Understanding: Investigation 1

1. $R(A)$
   $T(A)$
2. Always
   Sometimes
   Never
   Sometimes
   Always
   Always
3. M' would be mapped back onto M by a rotation of $(360 - x)°$ clockwise around the origin.

Check for Understanding: Investigation 2

1. D, E
2. B
3. I would find the midpoints of $\overline{XX'}$ and $\overline{YY'}$. The line through these two midpoints is the line of reflection.

Check for Understanding: Investigation 3

1. A. III.
   B. V.
   C. VI.
   D. II.
   E. IV.
   F. V.
2. C
3. First find the midpoint of $\overline{AA'}$, which is $\left(\frac{-5+1}{2}, \frac{3+3}{2}\right)$ or $(-2, 3)$. The midpoint of $\overline{BB'}$ is $\left(\frac{-8+4}{2}, \frac{-3+-3}{2}\right)$ or $(-2, -3)$. The line that passes through $(-2, -3)$ and $(-2, 3)$ is $x = -2$, so $x = -2$ is the line of reflection.

Check for Understanding: Investigation 4

1. D
2. E
3. Yes, slopes of segments connecting the pre-image points to the center and the corresponding image points to the center are negative reciprocals. $\underline{AP} \perp \underline{RP}$, $\underline{BP} \perp \underline{SP}$, and $\underline{CP} \perp \underline{TP}$. Lengths of each segment from the center were equal. Lengths: $AP = RP$, $BP = SP$, and $CP = TP$.

   $AP = \sqrt{(-10-0)^2 + (1-[-2])^2} = \sqrt{109}$, $RP = \sqrt{(3-0)^2 + (8-[-2])^2} = \sqrt{109}$

   $BP = \sqrt{(-4-0)^2 + (5-[-2])^2} = \sqrt{65}$, $SP = \sqrt{(7-0)^2 + (2-[-2])^2} = \sqrt{65}$

   $CP = \sqrt{(-2-0)^2 + (2-[-2])^2} = \sqrt{20}$, $TP = \sqrt{(4-0)^2 + (0-[-2])^2} = \sqrt{20}$

Explore Transformations *(continued)*

Check for Understanding: Investigation 5

1. 8
2. A, B; C; A, B; A
3. B
4. For $\overline{WY}$ to be a line of symmetry, each side would have to map across $\overline{WY}$ onto one of the other sides. Because one side is longer than the other, that means $\angle YWX$ and $\angle YWZ$ are not congruent. This means that if $\overline{WX}$ were reflected across $\overline{WY}$, it wouldn't map onto $\overline{WZ}$, nor any of the other sides, so $\overline{WY}$ cannot be a line of symmetry.

Check for Understanding: Investigation 6

1. False
   True
   False
   True
2. $m\angle 1 = 50°$
   $m\angle 2 = 65°$
   $m\angle 3 = 65°$
   $m\angle 4 = 50°$
   $m\angle 5 = 65°$
   $m\angle 6 = 65°$
3. Yes, it is possible, but you would have to use another type of polygon to avoid having any gaps or overlaps in the tessellation. For example, an irregular quadrilateral can be used to fill the space between the trapezoids, thus making a row of trapezoid-quadrilaterals, then stacking those rows on top of each other.

Practice Exercises

1. B, D, E
2. B, C, E
3. B
4. $A'(4, -2)$
   $B'(5, -3)$
   $C'(4, -4)$
   $D'(3, -2)$
5. A
6. B
7. 5, 72°
8. C
9. B
10. B

Explore Transformations *(continued)*

Apply

What Path Should the Rover Take?

This problem has students determine the shortest path the rover can take to visit all three rocks and then travel to its ending point. Students will model the path using transformations.

At this point, students are not expected to use trigonometry to determine the angles through which the rover must rotate. Instead students can print out the map and measure the angles with a protractor. They also can import the map into the Dynamic Geometry Tool and use the tool to measure the angles.

The Apply problems are rich performance tasks that may have a variety of open-ended solutions. One sample student response for this problem can be found in the Model Lesson.

2.2 Investigate and Apply Congruence Definitions

Check for Understanding: Investigation 1

1. D
2. D
3. Yes. Any translation that maps $\overline{AB}$ into a position could be rotated back into the original position. The angle 180 degrees would do this.

Check for Understanding: Investigation 2

1. A. Always
   B. Never
   C. Sometimes
   D. Sometimes
2. rotation, P, 50°
3. 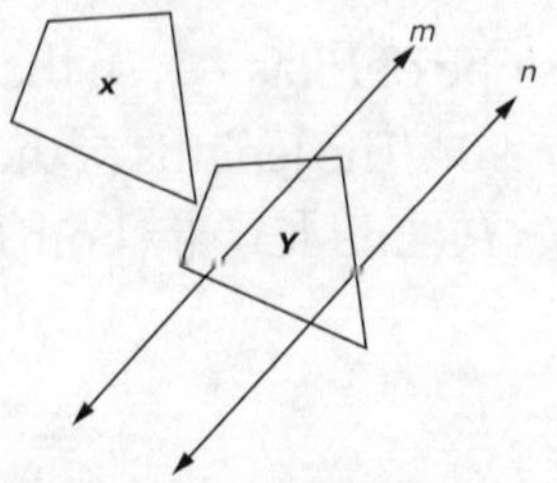

4. Measure the perpendicular distance between m and n. A translation twice that distance, in the direction of that is perpendicular to both lines, would be equivalent to the two reflections.

Check for Understanding: Investigation 3

1. C
2. $(y-2, x+1)$, $(-x-1, y-2)$, $(x+1, -y+2)$
3. No, $T_{\langle 1,-2\rangle}(R_{y=x}(x,y)) = (y+1, x-2)$ which does not equal $R_{y=x}(T_{\langle 1,-2\rangle}(x,y)) = (y-2, x+1)$.

   $T_{\langle 1,-2\rangle}(R_{y-axis}(x,y)) = (-x+1, y-2)$ which does not equal $R_{y-axis}(T_{\langle 1,-2\rangle}(x,y)) = (-x-1, y-2)$.

   And $T_{\langle 1,-2\rangle}(R_{x-axis}(x,y)) = (x+1, -y-2)$ which does not equal $R_{x-axis}(T_{\langle 1,-2\rangle}(x,y)) = (x+1, -y+2)$.

Check for Understanding: Investigation 4

1. C
2. B, C, D
3. 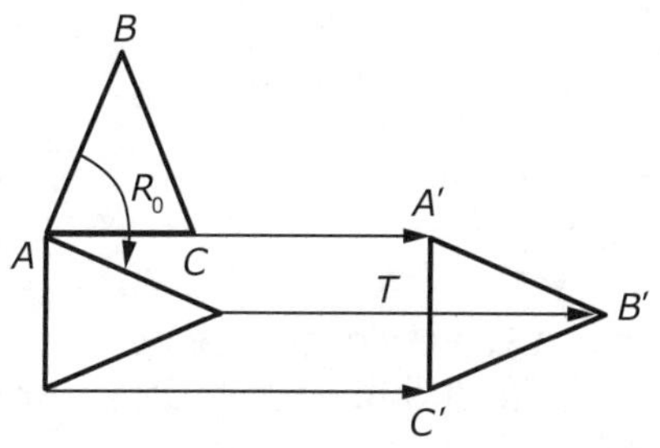

   $T(Ro_A(ABC)) = A'B'C'$

Investigate and Apply Congruence Definitions *(continued)*

Check for Understanding: Investigation 5

1. B, C, D, F

2. B

3.

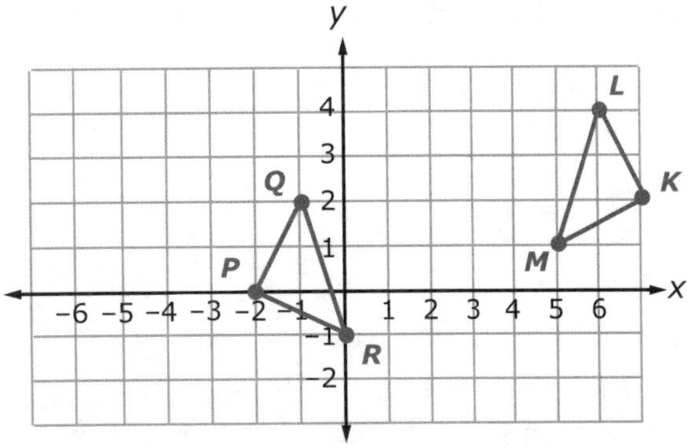

Yes. The two triangles are both right triangles, since the slope of $\overline{PQ}$ is 2 and the slope of $\overline{PR}$ is $-\frac{1}{2}$, so that $m\angle QPR = 90°$. The slope of $\overline{KL}$ is −2 and the slope of $\overline{KM}$ is $\frac{1}{2}$, so that $m\angle LKM = 90°$. The lengths of the hypotenuses $\overline{QR}$ and $\overline{LM}$ are both $\sqrt{10}$. Also, the lengths of the corresponding legs $\overline{PR}$ and $\overline{KM}$ are both $\sqrt{5}$. So the two triangles are congruent by the *HL* congruence criteria.

Practice Exercises

1. A, B, E, F

2. A

3. 1, −5

4. B, C

5. B, C, E

6. A, C, D

7. B

8. $W''(-3, 1)$
$X''(-1, 4)$
$Y''(-3, 3)$
$Z''(-5, 3)$

9. C

10. Yes

Apply

How Are Geometric Transformations Used to Design Textile Patterns?

In this problem, students make use of rigid motions to create a pattern that can be used for a sweater, fabric designs, or other articles of clothing. Students will research textile patterns, choose a design, and describe any transformations. Then students will write a sequence of rigid motions using function notation that describes their own pattern for an article of clothing.

In addition, consider the following research ideas to extend learning:

- How clothing patterns are created by using computers.
- Mechanical tools used to manufacture the articles of clothing.

The Apply problems are rich performance tasks that may have a variety of open-ended solutions. One sample student response for this problem can be found in the Model Lesson.

3.1 Explore Parallel and Perpendicular Lines

Check for Understanding: Investigation 1

1. C
2. would, would
3. $\frac{3}{5}, -\frac{5}{3}$
4. If 1st and 2nd Avenue are parallel to each other, then Arch and Market Streets are parallel, too, because the distance between parallel lines is constant. However, if they are not parallel to each other, then Arch and Market Streets are not parallel, either. The distance between the first pair of intersections could not equal the distance between the second pair if 1st and 2nd Avenue were not parallel to each other, but Arch and Market Streets were.

Check for Understanding: Investigation 2

1. top right
2. B
3. 1, 2
4. No, the pre-image and image of the existing line segment would not coincide, because the existing line segment would not be perpendicular to the new line segment. This is because the existing line segment is not parallel to the diagonal going from the top right corner to the bottom left corner. Only if it were would the pre-image and image of the existing line segment coincide after the reflection over the new line segment.

Check for Understanding: Investigation 3

1. A
2. does not, will
3. C, E
4. Two transformations are rotations of 90° and 270° counterclockwise. For the 90° rotation, point *A*'s image would be to the left and below its pre-image if the rotation's center is below point *A*; to the right and above if the center is above point *A*; and in the same place if the center is point *A*. For the 270° rotation, point *A*'s image would be to the right and below its pre-image if the rotation's center is below point *A*; to the left and above if the center is above point *A*; and in the same place if the center is point *A*.

Explore Parallel and Perpendicular Lines *(continued)*

Check for Understanding: Investigation 4

1. D
2. –4, –13
3. Ashton is correct that $\overleftrightarrow{AB}$ and $\overleftrightarrow{A'B'}$ are perpendicular, because the slope of $\overleftrightarrow{AB}$ is $\frac{4-(-4)}{6-(-2)} = \frac{8}{8} = 1$, while the slope of $\overleftrightarrow{A'B'}$ is $\frac{4-(-4)}{-6-2} = \frac{8}{-8} = -1$, and 1 and –1 are opposite reciprocals. Therefore, the two lines form a 90° angle. However, this will not be true for all lines. For example, the image of a line passing through (1, 1) and (2, 4) would pass through (–1, 1) and (–2, 4). In this case, the slopes of the pre-image and image of the line would be $\frac{4-1}{2-1} = \frac{3}{1} = 3$ and $\frac{4-1}{-2-(-1)} = \frac{3}{-1} = -3$, and these are not opposite reciprocals.

Practice Exercises

1. C
2. B, C
3. B
4. $m = \frac{1}{4}, b = -3$
5. B
6. B, D, E
7. C
8. $-\frac{12}{5}, \frac{507}{5}$
9. A
10. C

Apply

Where Are the Parallel and Perpendicular Folds in an Origami Design?

This problem allows students to find a set of instructions for a simple origami design online, and then follow the instructions to create a three-dimensional sculpture. They will then unfold the paper, mark all perpendicular and parallel lines, and explain how they know that the lines are perpendicular or parallel.

Instructions for cutting a square sheet of paper from a standard rectangular sheet of paper:

- Turn the paper so that the shorter sides are on the top and bottom.
- Hold the top right corner, and fold the top side down and to the left so that the top side aligns exactly on top of the left side as shown.

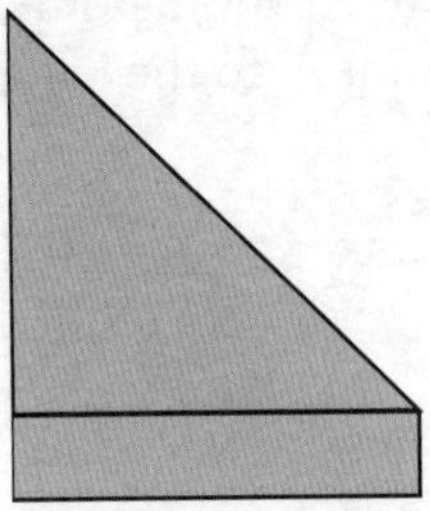

- Cut off the rectangle at the bottom of the paper. Then, unfold the remaining piece.

If teachers would like to provide students with a set of origami instructions, they can provide students with the Hands-On Activity: Origami Instructions.

The Apply problems are rich performance tasks that may have a variety of open-ended solutions. One sample student response for this problem can be found in the Model Lesson.

3.2 Prove Theorems about Lines and Angles

Check for Understanding: Investigation 1

1. is, not all
2. 180°, $m\angle ABC + m\angle CBD = 180°$
3. The structure should be modified. In the top two nodes, Kathy should write the given statements, which are "$\angle 1$ is a straight angle" and "$\angle 2$ is a straight angle." In the nodes under those, she should write "$m\angle 1 = 180°$" and "$m\angle 2 = 180°$" by the definition of straight angles. In the next node down, she should write "$m\angle 1 = m\angle 2$" by the transitive property of equality. Finally, in the node under that, she should write "$\angle 1 \cong \angle 2$" by the definition of congruent angles. Kathy should now delete the last arrow and node.

Check for Understanding: Investigation 2

1. B
2. corresponding angels theorem, substitution property of equality
3. By the alternate interior angles theorem, $\angle 4 \cong \angle 5$, so by the definition of congruent angles, $m\angle 4 = m\angle 5$. By the linear pair property, $\angle 5$ and $\angle 7$ are supplementary, so $m\angle 5 + m\angle 7 = 180°$ by the definition of supplementary angles. By the substitution property of equality, $m\angle 4 + m\angle 7 = 180°$, and again by the substitution property of equality, $m\angle 1 + m\angle 7 = 180°$. Therefore, $\angle 1$ is supplementary to $\angle 7$ by the definition of supplementary angles.

Check for Understanding: Investigation 3

1. B, C
2. C
3. Yes, you can. By the definition of supplementary angles, $m\angle 2 + m\angle 8 = 180°$. By the linear pair theorem, $\angle 6$ is supplementary to $\angle 8$, so by the definition of supplementary angles, $m\angle 6 + m\angle 8 = 180°$. By the substitution property of equality, $m\angle 2 + m\angle 8 = m\angle 6 + m\angle 8$, and by the subtraction property of equality, $m\angle 2 = m\angle 6$. By the definition of congruent angles, $\angle 2 \cong \angle 6$, so by the converse of the corresponding angles theorem, $m \parallel n$.

Check for Understanding: Investigation 4

1. B
2. reason, third
3. Yes, he did. He cannot state that $m\angle 1 + m\angle 3 = 180°$ and $m\angle 2 + m\angle 3 = 180°$ by the definition of straight angles. While it is true that the measure of a straight angle is 180°, he cannot assume that one of the sides of either $\angle 1$ or $\angle 2$ and one of the sides of $\angle 3$ form a line. Therefore, with the information Quill was given, it was not possible for him to correctly complete his proof.

Prove Theorems about Lines and Angles *(continued)*

Check for Understanding: Investigation 5

1. B, E

2. A

3. Both are correct. The coordinates for F' and G' will be those Stella's teacher gave if $\overleftrightarrow{FG}$ is rotated 90° counterclockwise or 270° clockwise about the origin. They will be those Stella gave if $\overleftrightarrow{FG}$ is rotated 270° counterclockwise or 90° clockwise. The slope of $\overleftrightarrow{FG}$ is $\frac{y_2 - y_1}{x_2 - x_1}$. The slope of $\overleftrightarrow{F'G'}$ can be found as $\frac{x_2 - x_1}{-y_2 - (-y_1)} = \frac{x_2 - x_1}{-(y_2 - y_1)} = -\frac{x_2 - x_1}{y_2 - y_1}$ or $\frac{-x_2 - (-x_1)}{y_2 - y_1} = \frac{-(x_2 - x_1)}{y_2 - y_1} = -\frac{x_2 - x_1}{y_2 - y_1}$. Either way, the slope of $\overleftrightarrow{F'G'}$ is the opposite reciprocal of the slope of $\overleftrightarrow{FG}$.

Practice Exercises

1. $m\angle 1 = 112°$
$m\angle 2 = 68°$
$m\angle 3 = 68°$

2. 5, 3, 180

3. 29°

4. 90°, 2, –2

5. 16

6. A

7. C

8. alternate, 2, 2, congruent

9. C

10. 180°, linear, vertical

Apply

How Are Optical Illusions Created?

Students will look at two optical illusions and decide whether they are parallel. Then, they will decide if the lines are parallel by using theorems learned in the unit. They research how the illusions work, and use their knowledge to create their own optical illusion. They provide a mathematical argument about why their illusion works.

The Apply problems are rich performance tasks that may have a variety of open-ended solutions. One sample student response for this problem can be found in the Model Lesson.

4.1 Prove Congruence Theorems

Check for Understanding: Investigation 1

1. C
2. 88°, 56°
3. No, Ed did not prove this, but there is nothing wrong with his proof. He did prove successfully that $m\angle 1 + m\angle 2 + m\angle 3 + m\angle 4 + m\angle 5 + m\angle 6 = 360°$. The problem is that neither $\angle 5$ nor $\angle 6$ are interior angles of $\triangle PQR$. Both $\angle 1$ and $\angle 4$ are, and the larger angle comprised of $\angle 2$ and $\angle 3$ is. The sum of the measures of these angles is 180°. The additional 180° in the last statement of Ed's proof come from the straight angle that is comprised of $\angle 5$ and $\angle 6$.

Check for Understanding: Investigation 2

1. B
2. AAS
3. To apply the HL triangle congruence theorem, Phillipe has to know that both triangles are right triangles and that two sides are congruent. The Pythagorean theorem can then be applied to show that the third side is also congruent, which means you can apply the SSS triangle congruence theorem.

Check for Understanding: Investigation 3

1. $\angle B$ or $\angle ABC$
2. B, E
3. A
4. Dale can't, but Felipe can. First, Felipe could prove that $\angle P \cong \angle Q \cong \angle R$. Then he could state that $m\angle P = m\angle Q = m\angle R$ by the definition of congruent angles. Next he could state $m\angle P + m\angle Q + m\angle R = 180°$ by the triangle sum theorem, and after that, $m\angle P + m\angle P + m\angle P = 180°$, $m\angle Q + m\angle Q + m\angle Q = 180°$, and $m\angle R + m\angle R + m\angle R = 180°$ by the substitution property of equality. Then $3(m\angle P) = 180°$, $3(m\angle Q) = 180°$, and $3(m\angle R) = 180°$ by addition. Finally, $m\angle P = 60°$, $m\angle Q = 60°$, and $m\angle R = 60°$ by the division property of equality.

Check for Understanding: Investigation 4

1. D, J
2. B
3. First, Genevieve could let point M be the midpoint of $\overline{RT}$ in the diagram and draw $\overline{SM}$. By the definition of a midpoint, $RM = TM$, and by the reflexive property, $SM = SM$. By the definition of congruence, $\overline{RS} \cong \overline{TS}$, $\overline{RM} \cong \overline{TM}$, and $\overline{SM} \cong \overline{SM}$, so by the SSS triangle congruence theorem, $\triangle RSM \cong \triangle TSM$. This means $\angle RMS \cong \angle TMS$ by corresponding parts of congruent triangles. Now Genevieve will be able to prove the converse of the perpendicular bisector theorem by showing that point S lies on the perpendicular bisector of $\overline{RT}$.

Prove Congruence Theorems *(continued)*

Practice Exercises

1. 26.5
2. 60
3. 7
4. 120
5. C, E
6. 54, 63
7. D
8. C
9. FAS, HL
10. A

Apply

How Will Frank Lloyd Wright Inspire You?

This problem provides students opportunities to apply the triangle congruency theorems within their own art. Students find the least possible number of measurements to prove triangles are congruent.

- If research is not an option, students could look at the introductory image of one of Frank Lloyd Wright's designs for inspiration.
- As an extension, students can generalize the number of measurements needed to prove whether any two triangles are congruent.

The Apply problems are rich performance tasks that may have a variety of open-ended solutions. One sample student response for this problem can be found in the Model Lesson.

4.2 Construct Special Triangles and Angles

Check for Understanding: Investigation 1

1. D
2. is, is
3. Yes, Kendrick is correct. In the figure, each line segment connecting the center of the circle to one of the hexagon's vertices is a radius of the circle. This means that $\triangle ROT$, $\triangle ROS$, and $\triangle SOT$ are congruent isosceles triangles, so the triangles' base angles must all be congruent to each other. Since $\angle SRT$, $\angle RST$, and $\angle RTS$ each consist of two of the base angles, these three larger angles must all be congruent to each other. Therefore, the measure of each of these three angles must be 60°, so $\triangle RST$ must be equilateral.

Check for Understanding: Investigation 2

1. D
2. 3, 3
3. Even though the student mixed up steps 5 and 7, each of the sides of the triangle the student constructed is still congruent to one of the sides of $\triangle GHK$. Therefore, by the SSS triangle congruence theorem, the triangle is congruent to $\triangle GHK$. If the student hadn't mixed up the steps, the triangle could have been mapped onto $\triangle GHK$ with translations and a rotation at most. Since the student mixed up the steps, a reflection will also be needed.

Check for Understanding: Investigation 3

1. C
2. greater than
3. Yes, she is correct. Suppose $\overline{OS}$ were drawn. Since $\overline{OR}$, $\overline{OS}$, and $\overline{OT}$ are all radii of the circle, both $\triangle ROS$ and $\triangle SOT$ are isosceles triangles. So their base angles would be congruent by the isosceles triangle theorem. If the base angles of $\triangle ROS$ measure $x°$, and those of $\triangle SOT$ measure $y°$, then $m\angle SRT = x°$, $m\angle RST = (x + y)°$, and $m\angle STR = y°$. By the triangle sum theorem, $x + (x + y) + y = 180$. This simplifies to $2x + 2y = 180$, which is equivalent to $x + y = 90$, so $m\angle RST = 90°$. Thus, $\overline{RS} \perp \overline{ST}$ by the definition of perpendicular lines.

Check for Understanding: Investigation 4

1. 45°, 45°
2. does, does not, does
3. Suppose Fiona were to call the point where the first arc she drew crosses the ray P. Assuming that the point of intersection of the two arcs she drew is L, then, if drawn, $\triangle LMP$ would be an equilateral triangle, because Fiona never changed her compass span. The triangle's interior angles would measure 60°, so this is the measure of the angle Fiona has constructed.

Construct Special Triangles and Angles *(continued)*

Practice Exercises

1. D, A, C, B
2. C
3. A, B, E
4. B, C, F
5. D
6. A
7. B, C
8. DF, DE

Apply

How Can You Copy a Design with Only a Straightedge and Compass?

This problem allows students to demonstrate their ability to make geometric constructions. Students will accurately re-create a design containing several different sizes of triangles in different orientations using only a straightedge and compass.

- Encourage students to write down the steps they take as they go along, instead of listing them from memory after they are finished.
- The Evaluation Criteria provides some guidance for the students on how to construct the image. Advanced students may wish to try it entirely on their own and look at the Evaluation Criteria when they are finished.

Note that to construct the center of the triangle, students can either construct the perpendicular bisector of each side of the triangle or the angle bisector of each angle of the triangle. For an equilateral triangle, the intersection of the three perpendicular bisectors is the same as the intersection of the three angle bisectors.

The Apply problems are rich performance tasks that may have a variety of open-ended solutions. One sample student response for this problem can be found in the Model Lesson.

5.1 Explore Similarity and Dilation

Check for Understanding: Investigation 1

1. A, G, H
2. 72, $\frac{1}{3}$
3. It appears that the rectangles are congruent. If so, the scale factors of the two dilations must be reciprocals. Since the second dilation had a center at vertex Z' of rectangle $W'X'Y'Z'$, Z' was in the same location as Z'', which means that Z' was farther away from P than Z. Thus, the scale factor of the first dilation was greater than 1, and that of the second was less than 1. It looks like Z'' is about 5 times as far from P as Z, so the scale factor of the first dilation could have been 5, and that of the second could have been $\frac{1}{5}$.

Check for Understanding: Investigation 2

1. D, F, G
2. (–10, –15), (5, –10), (15, –5), (20, 10)
3. First, Bo could translate both $\triangle XYZ$ and P down and to the left 2 units. The vertices of $\triangle XYZ$ would then be $X(-6, 0)$, $Y(-4, 2)$, and $Z(2, -6)$, while P would be at the origin. Then, he could use the rule for a dilation by a scale factor of k centered at the origin, which is $D_k(x, y) = (kx, ky)$, to dilate $\triangle XYZ$. The vertices of $\triangle X'Y'Z'$ would be $X'(-3, 0)$, $Y'(-2, 1)$, and $Z'(1, -3)$. Finally, he could translate both $\triangle X'Y'Z'$ and P up and to the right 2 units. The vertices of $\triangle X'Y'Z'$ would then be $X'(-1, 2)$, $Y'(0, 3)$, and $Z'(3, -1)$.

Check for Understanding: Investigation 3

1. 5, 16
2. D
3. Yes, he could have. If Jorge would have first reflected $\triangle LMN$ across $\overleftrightarrow{PQ}$, L' would have been in the same place as X. He could have then used L' as the center of dilation when dilating $\triangle L'M'N'$, and the end result would have been the same as with the combined transformation $R_{\overleftrightarrow{PQ}}\left(D_{1.5}(\triangle LMN)\right)$. In this case, his composition of transformations would have been $D_{1.5}\left(R_{\overleftrightarrow{PQ}}(\triangle LMN)\right)$.

Explore Similarity and Dilation *(continued)*

Practice Exercises

1. –9, 1

2. $\frac{8}{20}$, 0.4

3. (4, –4)

4. D, E, F

5.

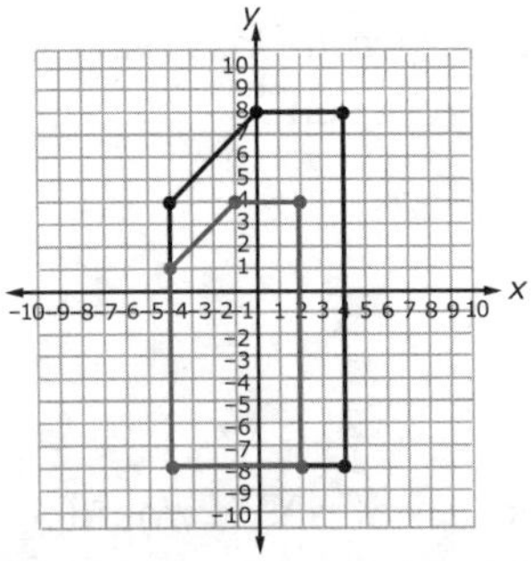

6. 0.269

7. D

8. E

9. $a = 4.31$
$b = 4.66$
$m\angle C = 99.2$
$m\angle D = 149$

10. Point B': (0, 9)
Point C': (7.5, –7.5)

Apply

Where Is the Third Vertex?

In this problem, students use coordinates to locate the vertex of a triangle in a mural. The coordinates of the image are given after a combined transformation, and the students must work backward to determine all possible locations of the vertex.

- Encourage students to try different methods, including graphing paper with cutouts and the Graphing Calculator.
- The difficulty level of the problem can be modified as needed by changing the requirements. For example, to simplify the problem, you could specify the types of transformations, or to make it more challenging, the given side could be any side of the triangle.

The Apply problems are rich performance tasks that may have a variety of open-ended solutions. One sample student response for this problem can be found in the Model Lesson.

5.2 Prove Similarity Theorems

Check for Understanding: Investigation 1

1. C
2. AA
3. Yes, Juan is correct. He can dilate $\triangle RST$ by a scale factor of $k = \frac{XY}{RS}$ and with a center of dilation at S to get $\triangle R'ST'$. Then he will know that $R'S = XY$ and $R'T' = XZ$. Since angle measures are unchanged under dilation, $\triangle R'ST'$ will contain a right angle and will be a right triangle. Therefore, by the HL congruence theorem, $\triangle R'ST' \cong \triangle XYZ$. This means that there is a composition of rigid motion transformations F such that $F(\triangle R'ST') = \triangle XYZ$. It follows that $F \circ D_k(\triangle RST) = \triangle XYZ$, so by the definition of similarity, $\triangle RST \sim \triangle XYZ$.

Check for Understanding: Investigation 2

1. 15, 28, 21
2. D, F
3. do not, is
4. By the definition of a midpoint, $NR = RP = \frac{1}{2}NP$, $PS = SQ = \frac{1}{2}PQ$, and $NT = TQ = \frac{1}{2}NQ$. Also, by the triangle midsegment theorem, $ST = \frac{1}{2}NP$, $RT = \frac{1}{2}PQ$, and $RS = \frac{1}{2}NQ$. By substitution, $NR = RP = ST$, $PS = SQ = RT$, and $NT = TQ = RS$, so by the definition of congruence, $\overline{NR} \cong \overline{RP} \cong \overline{ST}$, $\overline{PS} \cong \overline{SQ} \cong \overline{RT}$, and $\overline{NT} \cong \overline{TQ} \cong \overline{RS}$. By the SSS triangle congruence theorem, $\triangle NRT \cong \triangle RPS \cong \triangle STR \cong \triangle TSQ$, and by the SSS similarity criteria for triangles, $\triangle NRT \sim \triangle RPS \sim \triangle STR \sim \triangle TSQ \sim \triangle NPQ$. Therefore, Valerie is correct.

Check for Understanding: Investigation 3

1.

	Length of Shorter Leg	Length of Longer Leg	Length of Hypotenuse
Harry's Image	*aa*	*ab*	*ac*
Rosa's Image	*ab*	*bb*	*bc*
Finn's Image	*ac*	*bc*	*cc*

2. D
3. When the three students dilated their right triangles, the three images produced were each similar to the pre-images. Since the pre-images were congruent, the images were also similar to each other. This means they were each right triangles, so when Rosa and Harry put their triangles together as shown in the figure, it produced a larger right triangle with sides of lengths ac, bc, and $aa + bb$. Since Finn's image has leg lengths of ac and bc and a hypotenuse length of cc, $aa + bb = cc$, or $a^2 + b^2 = c^2$.

Prove Similarity Theorems *(continued)*

Check for Understanding: Investigation 4

1. $\angle DBE$
2. D
3. less than
4. Since $CD = 56$ in., when Norbert found the baton's shadow to measure 8 in., the length of $\overline{DE}$ was $56 + 64 = 120$ in. Also, $\triangle DBE \sim \triangle FGH$, so the corresponding parts of these two triangles are proportional. This means that the proportion $\frac{8}{14} = \frac{120}{x}$ can be set up, where x is the height of the tent in inches. Solving for x shows that the tent's height is 210 in., or $\frac{210}{12} = 17.5$ ft.

Practice Exercises

1. B, C, F, H
2. A
3. C
4. $BE = 4$
   $ED = 5$
   $DA = 2.5$
5. C
6. C
7. B
8. C
9. 6
10. 240

Apply

How Tall Are Those Towering Objects?

This problem allows students to demonstrate their ability to set up and solve proportions. Students will use the shadow method to determine the height of a tall object.

- You may want to have students watch the video first, and then work in pairs outside.
- For long shadows, each student can hold one end of the tape measure. It is also helpful to have students work in pairs so one can measure the shadow of the other. If a student is working independently, it may help to stand so that the shadow ends at a place that is easy to measure to.
- As an extension, students can compare their results for two or more methods of indirect measurement.

The Apply problems are rich performance tasks that may have a variety of open-ended solutions. One sample student response for this problem can be found in the Model Lesson.

5.3 Apply Similarity Theorems

Check for Understanding: Investigation 1

1. A
2. can, cannot
3. 32, 35
4. A, E
5. Each prism has five faces, two of which are equilateral triangles, and three of which are rectangles. Since all equilateral triangles have three interior angles measuring 60° each, the bases of one prism must be similar to those of the other. However, the rectangular faces do not have to be similar, because the bases of one prism do not have to be congruent to those of the other. Therefore, the two prisms do not have to be similar to each other.

Check for Understanding: Investigation 2

1. 6.8, 1.7, 8.5
2. A, E
3. volume
4. Since the shape's volume increased by 2,352 cm^3, its volume after the dilation was $336 + 2,353 = 2,688\ cm^3$. The ratio of the shape's volume after the dilation to its volume before the dilation is $\frac{2,688\ cm^3}{336\ cm^3} = 8$, so the scale factor must have been $\sqrt[3]{8} = 2$. This means that the dilation caused the shape's surface area to increase from 292 cm^2 to $292 \cdot 2^2 = 1,168\ cm^2$. Therefore, the shape's surface area increased by $1,168 - 292 = 876\ cm^2$.

Practice Exercises

1. C
2. 8
3. 16
4. 9
5. 8
6. D
7. A
8. 6.75, 3, 6
9. C
10. 1.2

Apply Similarity Theorems *(continued)*

Apply

How Much Bigger Is That Box?

In this problem, students compare dimensions of standard cardboard boxes.

- Students who do not have Internet access at home may need to complete the research at school.
- You may want to bring cardboard boxes in for students to measure. If students are having difficulty finding two boxes that are similar, have them look at one box and make a list of the dimensions of possible similar boxes, then look for a box that has one of those dimensions. It may be easier to focus on boxes with square bases.
- As an extension, consider the extra material used for the flaps of the box, and determine how that relates to the area of cardboard used and the price of the box.

The Apply problems are rich performance tasks that may have a variety of open-ended solutions. One sample student response for this problem can be found in the Model Lesson.

6.1 Investigate Right Triangle Trigonometry

Check for Understanding: Investigation 1

1. 8, 8, $16+8\sqrt{2}$ or $8\sqrt{2}+16$, 32
2. D, E
3. Since the sum of the measures of the angles in a triangle is 180°, Waldron Trail and Rim Trail must make a $180°-90°-45°=45°$ angle, while River Trail and Rim Trail must make a $180°-90°-30°=60°$ angle. This means that Waldron Trail, Beamer Trail, and Rim Trail make up a 45-45-90 right triangle, while Hermit Trail, River Trail, and Rim Trail make up a 30-60-90 right triangle. Therefore, Beamer Trail's length is 4 km, and Rim Trail's length is $4\sqrt{2}$ km. Also, River Trail's length is $\frac{4\sqrt{2}}{2}=2\sqrt{2}$ km, and Hermit Trail's length is $2\sqrt{2}\cdot\sqrt{3}=2\sqrt{6}$ km.

Check for Understanding: Investigation 2

1. A. V.
   B. IV.
   C. I.
   D. VI.
   E. III.
   F. II.
2. D
3. 0.26
4. The sine of an acute angle in a right triangle is $\frac{\text{length of opposite leg}}{\text{length of hypotenuse}}$, while the cosine of the angle is $\frac{\text{length of adjacent leg}}{\text{length of hypotenuse}}$. Since $\frac{35}{37}$ was given for the sine of an angle, the leg opposite the angle in a right triangle is 35 units long, and the hypotenuse is 37 units long. According to the Pythagorean theorem, $a^2+35^2=37^2$, and solving for a gives $a=12$. This means the leg adjacent to the angle is 12 units long, so the cosine of the angle is $\frac{12}{37}$.

Check for Understanding: Investigation 3

1. C
2. 48.9°, 41.1°
3. After t minutes of walking north, Jon had walked $65t$ meters, and after t minutes of walking east, Kelly had walked $80t$ meters. This means the ratio of the distance walked by Jon to that walked by Kelly is $\frac{65t}{80t}=\frac{13}{16}$, no matter what the value of t is, so it isn't necessary to know the number of minutes Jon and Kelly had been walking. Since $\tan^{-1}\left(\frac{13}{16}\right)\approx 39.1°$, the angle bearing at which Kelly walked back to Jon is about $90°-39.1°\approx 50.9°$ west of north.

Investigate Right Triangle Trigonometry *(continued)*

Check for Understanding: Investigation 4

1. B, G
2. 107°, 14.34
3. The area given by Simon's formula should be compared to the area given by the formula Area = $\frac{1}{2}ab\sin C$. Suppose two sides of a triangle measure 5 and 12 units. If the included angle is 20°, both formulas give an area of about 10.26 units^2. If the angle is 140°, both formulas give an area of about 19.28 units^2. If the angle is 90°, Area = $\frac{1}{2}ab\sin C$ gives an area of 30 units^2, and Simon's formula gives an error, so Simon's formula doesn't work for right angles. This is because, according to the calculator, you can't find tan90°.

Practice Exercises

1. E
2. D
3. 60
4. $5\sqrt{3}+5$
5. C
6. F, D, C
7. 1
8. 0.8
9. 12.9
10. 16.4
11. B
12. B

Apply

How Will You Design a Ramp?

This problem allows students to apply their understanding of trigonometric ratios to solve a real-world problem. Students will solve right triangles to find angles and side lengths of an ADA-approved ramp.

- If some students will be working on this problem at home without Internet access, you may want to have them complete the research at school, or provide the following information about ADA guidelines:
  - The maximum slope of the ramp is 1 : 12.
  - The maximum rise for any part of a ramp is 30 in.
  - The minimum width is 36 in.
- Students can confirm their answers by creating scale drawings of the ramp in the Dynamic Geometry Tool or on graph paper, and measuring the unknown quantities.
- As an extension, students can measure the rise and run of a ramp in their school, and determine the angle of elevation.

The Apply problems are rich performance tasks that may have a variety of open-ended solutions. One sample student response for this problem can be found in the Model Lesson.

6.2 Investigate Laws of Sines and Cosines

Check for Understanding: Investigation 1

1. B
2. 48°, 47.6, 62.6
3. A, D
4. Suppose you have $\triangle ABC$ with a side of length a opposite $\angle A$, a side of length b opposite $\angle B$, and a side of length c opposite $\angle C$. The formula for the area of the triangle is Area $= \frac{1}{2}bc \sin A$, Area $= \frac{1}{2}ac \sin B$, or Area $= \frac{1}{2}ab \sin C$. This means $\frac{1}{2}bc \sin A = \frac{1}{2}ac \sin B = \frac{1}{2}ab \sin C$. After multiplying by 2, the result is $bc \sin A = ac \sin B = ab \sin C$. Then divide by abc to get $\frac{\sin A}{a} = \frac{\sin B}{b} = \frac{\sin C}{c}$, which is the law of sines.

Check for Understanding: Investigation 2

1. A
2. D, E
3. 82.8°
4. In the figure, $\triangle ABC$, which has side lengths of a, b, and c, is the acute triangle, and $\triangle ABD$ is a right triangle. The lengths of the legs of $\triangle ABD$ are $a \sin C$ and $b - a \cos C$, and the length of the hypotenuse is c. According to the Pythagorean theorem, $(a \sin C)^2 + (b - a \cos C)^2 = c^2$, and expanding the equation results in $a^2 \sin^2 C + b^2 - 2ab \cos C + a^2 \cos^2 C = c^2$. The symmetric property, along with rearranging and factoring, gives $c^2 = a^2(\cos^2 C + \sin^2 C) + b^2 - 2ab \cos C$. Because $\cos^2 C + \sin^2 C = 1$, the equation can be written as $c^2 = a^2 + b^2 - 2ab \cos C$, which is the law of cosines.

Check for Understanding: Investigation 3

1. C, H
2.

$m\angle A$	a	c	Ambiguous Case	Non-Ambiguous Case
26°	29	19		√
117°	21	33		√
41°	19	25	√	
140°	17	42		√
38°	36	23		√
15°	27	39	√	

For an ambiguous case, $m\angle A$ must be less than 90°, because if it were not, $\angle C$ could only be acute, and not obtuse. Also, a must be less than c, or else there would be only one possible value of b, given that $m\angle A$ must remain constant. In addition, since the smallest possible value of a would occur when $\angle ACB$ is a right angle, or when $a = c \sin A$, a must be greater than $c \sin A$. If it were not, the only possible triangle would be a right triangle.

Investigate Laws of Sines and Cosines *(continued)*

Check for Understanding: Investigation 4

1. 82.7°, 66.3°, 35.3°, 113.7°
2. B
3. Dirk's mistake was that he didn't realize that this is not an ambiguous case of the law of sines. One way he could have known this before doing his calculations would have been by noticing that $x > z$. Only if $x < z$ could this be an ambiguous case. After doing his calculations, he could have also noticed that if $m\angle X = 35°$ and $m\angle Z = 148.2°$, then $m\angle X + m\angle Z = 183.2°$. This is impossible, because the sum of the interior angles of a triangle is 180°.

Check for Understanding: Investigation 5

1. B
2. A, B
3. Let the isosceles triangle be $\triangle XYZ$, with $YZ = x$, $XZ = y$, and $XY = z$. If $y = 12$ cm and $m\angle X = 35°$, then $m\angle Z = 35°$, too, because the base angles of an isosceles triangle are congruent. This means that by the triangle sum theorem, $m\angle Y = 180° - 35° - 35° = 110°$. Using the law of sines produces the equation $\frac{\sin 110°}{12} = \frac{\sin 35°}{x}$, and solving for x gives $x \approx 7.3$ cm. Since the two non-base sides of the triangle are congruent, $x = z$, so $z \approx 7.3$ cm as well.

Practice Exercises

1. 23.9, 2.8
2. A, B
3. A, C
4. 3.15
5. DNE
6. 75.5
7. 322.8
8. Pythagorean
9. 86
10. 2.7

Apply

How Far Apart Are the Stars?

In this problem, students will use both the law of sines and the law of cosines to solve a diagram of the stars in Orion's belt.

- If some students will be working on this problem at home without Internet access, you may want to have them watch the video and complete the research at school.
- Measurements and solutions will vary according to the information students find while researching. For this reason, this problem provides an opportunity to discuss the use of credible resources when performing research.
- You can also use this problem to prompt a discussion of rounding errors and why answers should not be rounded until the last step of a calculation when possible.

The Apply problems are rich performance tasks that may have a variety of open-ended solutions. One sample student response for this problem can be found in the Model Lesson.

7.1 Investigate Circles and Parts of Circles

Check for Understanding: Investigation 1

1.

Combination	Inside Circle	On Circle	Outside Circle
chord and chord	√	√	
secant and secant	√	√	√
chord and secant	√	√	
tangent and tangent			√
chord and tangent		√	
secant and tangent		√	√

2. D
3. If Martin draws the chord first, the chord and tangent could intersect at two possible points, which would be the two endpoints of the chord. If Keiko draws the tangent first, the chord and tangent could intersect at only one possible point, which would be where the tangent touches the circle.

Check for Understanding: Investigation 2

1. double
2. C
3. 92°, 20°
4. First you could draw $\overline{OX}$. Since $\overline{OX}$ and $\overline{OY}$ would both be radii of circle O, they would be congruent. This means that $\triangle XOY$ would be isosceles. The measure of a central angle of a circle is equal to the measure of the intercepted arc, so $m\angle XOY$ would equal 122°. Because the base angles of an isosceles triangle are congruent, $m\angle OXY$ and $m\angle OYX$ would equal half of 180° – 122°, or 29°. Since $m\angle OYZ$ would equal 90°, $m\angle XYZ$ would equal 90° – 29° = 61°, or half of the measure of the arc intercepted by chord $\overline{XY}$.

Check for Understanding: Investigation 3

1. B, C, F, H
2. B
3. First draw $\overline{OW}$ and $\overline{OX}$, which will both be radii of length r. Since $\overline{OZ}$ is perpendicular to $\overline{WX}$, it bisects $\overline{WX}$, so $\overline{WY}$ and $\overline{XY}$ both have a length of $\frac{c}{2}$. Also, $\overline{OY}$ must have a length of $r - h$. Using either $\triangle OWY$ or $\triangle OXY$, the Pythagorean theorem gives $(r-h)^2 + \left(\frac{c}{2}\right)^2 = r^2$. Expanding the terms on the left gives $r^2 - 2hr + h^2 + \frac{c^2}{4} = r^2$, and simplifying gives $2hr = h^2 + \frac{c^2}{4}$; dividing both sides by $2h$ gives $r = \frac{h}{2} + \frac{c^2}{8h}$, which can be rewritten as $r = \frac{4h^2 + c^2}{8h}$.

Investigate Circles and Parts of Circles *(continued)*

Check for Understanding: Investigation 4

1. each, equal to
2. B, C, E, H
3. A
4. Yes, Condoleezza's claim is correct. A tangent line of a circle is perpendicular to a radius of the circle at the point of tangency, which means that it is also perpendicular to a diameter of the circle. Therefore, the two chords that Condoleezza drew were diameters, and the point at which two or more non-coinciding diameters of a circle intersect is the circle's center.

Check for Understanding: Investigation 5

1. C
2. 0.4, 2.5
3. If $\overline{XY}$ were drawn, its length would be $r_1 + r_2$. Suppose $\overline{XY}$ were the hypotenuse of a right triangle, with a horizontal leg that is $\frac{r_1 + r_2}{2}$ units long, and a vertical leg that has a length equal to the number of units Paige translated the image of $\odot X$ up. This would have to be a 30-60-90 right triangle, so Paige must have translated the image of $\odot X$ exactly $\frac{\sqrt{3}(r_1 + r_2)}{2}$ up. When she dilated the new image, the scale factor must have been $\frac{r_2}{r_1}$.

Practice Exercises

1. Diameter: E
   Radius: C
   Chord: A
   Tangent: D
   Secant: B
2. B
3. C
4. Diameter: E
   Radius: D, F
   Chord: C, G
   Secant: A
   Tangent: B
5. A, D, F
6. B, E
7. 2.5, 90
8. $m\angle DAB = 15$
   $m\angle ABC = 157.5$
   $m\angle BCD = 30$
9. 37

Investigate Circles and Parts of Circles *(continued)*

Apply

How Will You Redesign a Courtyard?

In this problem, students create a design for a courtyard that uses a certain amount of materials. The design must include a circular seating area and another circle that is as large as possible for the given space. The circle will have walkways that are chords of the circle. Students will need to compute areas and angles within the circles.

- Students can work in several teams as a competition to design the best courtyard. Have each team present their design and show how it meets the specifications. Then have the students vote on the best design.
- The difficulty level of the problem can be modified as needed by adding or removing some of the specifications.

The Apply problems are rich performance tasks that may have a variety of open-ended solutions. One sample student response for this problem can be found in the Model Lesson.

7.2 Investigate and Apply Area and Circumference Formulas

Check for Understanding: Investigation 1

1. C
2. more, small
3. There appear to be 60 squares entirely within the circle, and each has an area of $0.2 \cdot 0.2 = 0.04$ square units. This means the total area of all these squares is $60 \cdot 0.04 = 2.4$ square units. Also, there appear to be 28 squares partially within the circle, and each has an area of 0.04 square units. This means the total area of all these squares is $28 \cdot 0.04 = 1.12$ square units. Since the area of a unit circle is π, or about 3.14, square units, Juaquin must have gotten $\frac{3.14 - 2.4}{1.12} = \frac{0.74}{1.12} \approx 0.66 \approx 66\%$ for the percentage.

Check for Understanding: Investigation 2

1.

Scale Factor	Area of Resulting Image (square units)
3	28.27
6	113.10
8	201.06
12	452.39
14	615.75
18	1,017.88

2. D
3. The unit circle's area is π square units. Since a dilation of the unit circle by a scale factor of r gives a circle with an area of πr^2 square units, the areas of the other circles are $\pi(1.5)^2 = 2.25\pi$, $\pi(2)^2 = 4\pi$, and $\pi(2.5)^2 = 6.25\pi$ square units, respectively. This means the area of the inner shaded ring is $2.25\pi - \pi = 1.25\pi$ square units, and the area of the outer shaded ring is $6.25\pi - 4\pi = 2.25\pi$ square units. Therefore, the combined area of the shaded regions is $1.25\pi + 2.25\pi = 3.5\pi$ square units, or about 11.00 square units.

Investigate and Apply Area and Circumference Formulas *(continued)*

Check for Understanding: Investigation 3

1. $\frac{1}{5}$, $\frac{\pi}{5}$
2. A. III.
   B. VI.
   C. IV.
   D. II.
   E. I.
   F. V.
3. The sector's area is $\frac{x}{360}$ the area of the circle, or $\frac{x}{360}\pi r^2$ square units. The unshaded triangle in the figure has two sides of length r that have an included angle measuring $x°$. This means the triangle's area is $\frac{1}{2}r^2\sin(x°)$ square units. The area, A, of the shaded segment is the area of the sector minus the area of the triangle, so the formula is $A = \frac{x}{360}\pi r^2 - \frac{1}{2}r^2\sin(x°)$. Factoring out r^2 and simplifying gives the formula $A = r^2\left[\frac{\pi x - 180\sin(x°)}{360}\right]$.

Check for Understanding: Investigation 4

1. A
2. 4, 0.25
3. In a 180-sided inscribed or circumscribed regular polygon, the value of θ is $\frac{360°}{180} = 2°$, and the value of n is 180. This means that the perimeter of a 180-sided inscribed polygon is about $2\sin(1°) \cdot 180 \approx 6.2829$, while the perimeter of a 180-sided circumscribed polygon is about $2\tan(1°) \cdot 180 \approx 6.2838$. The perimeter of a unit circle is 2π, which, according to a calculator, is about 6.2832. Since $6.2832 - 6.2829 = 0.0003$ and $6.2838 - 6.2832 = 0.0006$, the perimeter of a 180-sided inscribed polygon gives a better approximation.

Check for Understanding: Investigation 5

1. A, D
2. D
3. The arc length formula is $a = \frac{\pi r n}{180}$, where a is the length of an arc intercepted by a central angle of $n°$ in a circle of radius r. When the length of the arc intercepted by the angle is equal to the length of the circle's radius, $a = r$, so $r = \frac{\pi r n}{180}$. Dividing both sides by r gives $1 = \frac{\pi n}{180}$, and solving for n gives $n = \frac{180}{\pi}$. Thus, there are exactly $\left(\frac{180}{\pi}\right)°$ in 1 radian. Since 360° are in a circle, dividing 360 by $\frac{180}{\pi}$ shows that 2π radians are in a circle.

Investigate and Apply Area and Circumference Formulas *(continued)*

Practice Exercises

1. D
2. E
3. 24,900
   26,100
4. D
5. C
6. 12.6, 79.1 or 79.2
7. 18
8. 128π
9. D
10. 2.00
    0.318
11. 114.6
12. C

Apply

What Is the Perfect Amount of Pizza Crust?

In this problem, students will compare the area of a sector with the area of an area bounded by two arcs. They may compute the areas for the entire circle first and then divide.

- Students who do not have Internet access at home may need to complete the research at school.
- If students have trouble finding the information online, have them sketch their own pizzas and determine the ratio of crust area to topping area based on their sketches.
- You may wish to use the student response panel to show the class answers. Connect with statistics by having students find the average.
- As an extension, ask students to write a formula in terms of the radius of the pizza and the radius of the area with toppings.

The Apply problems are rich performance tasks that may have a variety of open-ended solutions. One sample student response for this problem can be found in the Model Lesson.

7.3 Investigate and Interpret Circle Equations

Check for Understanding: Investigation 1

1. 3, 4, 5
2. C
3. do, do not
4. The diameter of the circle passes through the circle's center, and it is twice the length of a radius of the circle, so the center of the circle is the midpoint of the diameter. This means that according to the midpoint formula, the circle's center is at the point $\left(\frac{-8+22}{2}, \frac{9+(-3)}{2}\right) = (7, 3)$. Since a radius of the circle has endpoints at (7, 3) and (−8, 9), the distance formula shows that the radius' length is $\sqrt{(-8-7)^2 + (9-3)^2} = \sqrt{225+36} = \sqrt{261}$. Therefore, an equation that represents all points on the circle is $(x-7)^2 + (y-3)^2 = 261$.

Check for Understanding: Investigation 2

1. B
2. the same, the same
3. Seymour added the square of 6, which is 36, and the square of −12, which is 144, to both sides of the equation, and then he incorrectly identified the x-terms and y-terms as perfect square trinomials. He should have added the square of $\frac{6}{2} = 3$, which is 9, and the square of $\frac{-12}{2} = -6$, which is 36, to both sides. He then would have gotten $(x+3)^2 + (y-6)^2 = 61$ for the graphing form. The circle's center is at the point (−3, 6), and its radius is $\sqrt{61}$ units.

Investigate and Interpret Circle Equations *(continued)*

Practice Exercises

1. 5
(4, 6)
0 or 8

2. 2, −3, 8

3. 52

4. C, E

5. D

6. D

7. A

8. (−3, 5); 7

9. B, C, D

10.

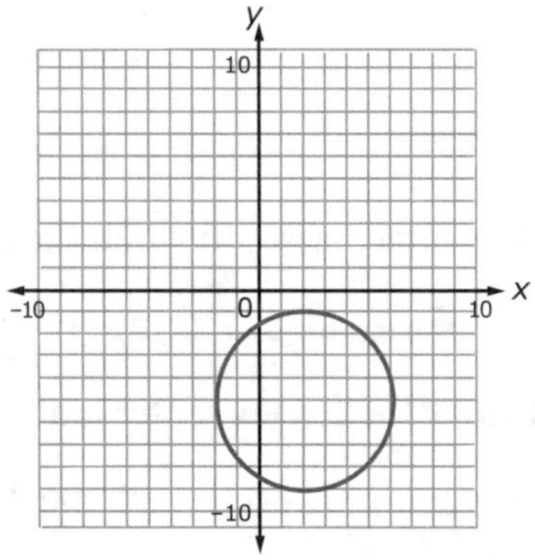

11. 128

Apply

Geometry Detective: What Is a Reuleaux Triangle?

In this problem, students explore the Reuleaux triangle and other Reuleaux polygons. They identify the equations of circles that are used in the construction and research possible applications.

- Students who do not have Internet access at home may need to complete the research and use the Dynamic Geometry Tool for the construction at school.
- As an extension, have students create rollers by cutting several copies of a Reuleaux polygon out of cardboard and gluing them together. Place two or more of these rollers on a table and set a book on top. As you roll the book back and forth, it should stay level, showing that the shape has a constant width. Mark the center on the triangle and ask students if it stays at the same level as the curve rotates. What does this mean about a bicycle with wheels that are Reuleaux triangles?

The Apply problems are rich performance tasks that may have a variety of open-ended solutions. One sample student response for this problem can be found in the Model Lesson.

8.1 Investigate Concurrency in Triangles

Check for Understanding: Investigation 1

1. D
2. A, B, F
3. Because $\odot O$ is inscribed in $\triangle XYZ$, point O is the incenter of $\triangle XYZ$. This means that point O is equidistant from $\overline{XY}$, $\overline{YZ}$, and $\overline{XZ}$, and the distance from point O to each of the triangle's sides is the radius of $\odot O$. If the radius of $\odot O$ is r, then the areas in square units of $\triangle XOY$, $\triangle XOZ$, and $\triangle YOZ$ are $\frac{1}{2}(3r)$, $\frac{1}{2}(4r)$, and $\frac{1}{2}(5r)$, respectively. The area of $\triangle XYZ$ is $\frac{1}{2}(3)(4) = 6$ square units, since it is a right triangle, so $\frac{1}{2}(3r) + \frac{1}{2}(4r) + \frac{1}{2}(5r) = 6$. Solving for r gives $r = 1$, so the area of $\odot O$ must be $A = \pi(1)^2 = \pi$ square units.

Check for Understanding: Investigation 2

1. D
2. B, D, E
3. Pedro can check if the slope of $\overline{YP}$ is equal to the slope of the median from vertex Y to $\overline{XZ}$, or if the slope of $\overline{ZP}$ is equal to the slope of the median from vertex Z to $\overline{XY}$. The slope of $\overline{YP}$ is $\frac{2-(-2)}{1-2} = -4$. Also, the midpoint of $\overline{XZ}$ is $\left(\frac{-2+3}{2}, \frac{5+3}{2}\right) = (0.5, 4)$, so the slope of the median from vertex Y to $\overline{XZ}$ is $\frac{4-(-2)}{0.5-2} = -4$. Therefore, this median and the one Pedro drew must intersect at point P. If two medians intersect at a point, the third median must also pass through that point, so Pedro's centroid must be correct.

Check for Understanding: Investigation 3

1. B
2. C, D, G
3. The medians intersect each of the sides of $\triangle PQR$ at their midpoints. This means that $PN = 1$, $NR = 1$, $RM = 1$, $MQ = 1$, $PL = \sqrt{2}$, and $LQ = \sqrt{2}$. Also, point X divides the medians into segments with lengths in the ratio 2:1. Thus, $PX = \frac{2\sqrt{5}}{3}$, $XM = \frac{\sqrt{5}}{3}$, $QX = \frac{2\sqrt{5}}{3}$, $XN = \frac{\sqrt{5}}{3}$, $RX = \frac{2\sqrt{2}}{3}$, and $XL = \frac{\sqrt{2}}{3}$. Therefore, the perimeters of $\triangle PXL$ and $\triangle QXL$ are $\frac{2\sqrt{5}}{3} + \frac{\sqrt{2}}{3} + \sqrt{2} = \frac{2\sqrt{5}+4\sqrt{2}}{3}$. The perimeters of $\triangle QXM$ and $\triangle PXN$ are $\frac{2\sqrt{5}}{3} + \frac{\sqrt{5}}{3} + 1 = \sqrt{5} + 1$. The perimeters of $\triangle RXM$ and $\triangle RXN$ are $\frac{2\sqrt{2}}{3} + \frac{\sqrt{5}}{3} + 1 = \frac{2\sqrt{2}+\sqrt{5}+3}{3}$.

Investigate Concurrency in Triangles *(continued)*

Check for Understanding: Investigation 4

1. D
2.

Point of Concurrency	Inside	On	Outside
Orthocenter of Acute Triangle	√		
Centroid of Acute Triangle	√		
Circumcenter of Acute Triangle	√		
Incenter of Acute Triangle	√		
Orthocenter of Right Triangle		√	
Centroid of Right Triangle	√		
Circumcenter of Right Triangle		√	
Incenter of Right Triangle	√		
Orthocenter of Obtuse Triangle			√
Centroid of Obtuse Triangle	√		
Circumcenter of Obtuse Triangle			√
Incenter of Obtuse Triangle	√		

3. Since $\triangle ABC$ is acute, its orthocenter, point *P*, must lie inside it. This means that point *C* must lie outside $\triangle ABP$, point *B* must lie outside $\triangle ACP$, and point *A* must lie outside $\triangle BCP$. Therefore, because $\triangle ABP$, $\triangle ACP$, and $\triangle BCP$ have orthocenters that lie outside them, these triangles must be obtuse.

Practice Exercises

1. C
2. D
3. D
4. 60
5. B
6. E
7. E
8. E
9. B
10. 20.6

Apply

Where Should the New Hospital Be Located?

This problem begins with a video about selecting a location for a new hospital. Students have the opportunity to demonstrate their knowledge of triangle centers to choose the best one and construct it. Then students extend their understanding to select a location that lies on an existing road. If you assign this problem for homework, note that students who do not have access to the Internet at home will need to watch the video Hospital Construction during class time. Students can print the map and construct the location using a compass and straightedge instead of the Geometry Tool. As an extension, have students present their locations to the class and defend their reasoning. Then hold a vote on which location is the best choice.

The Apply problems are rich performance tasks that may have a variety of open-ended solutions. One sample student response for this problem can be found in the Model Lesson.

9.1 Construct and Explore Polygons

Check for Understanding: Investigation 1

1. D, H
2. C
3. If a radius were drawn to each vertex of the pentagon, five congruent isosceles triangles would be formed. The vertex angle of each triangle would also be a central angle of the circle and measure $\frac{360°}{5} = 72°$. Each central angle would have a corresponding inscribed angle that would intercept the same arc, and these inscribed angles would form the points of the star. Thus, since the measures of the angles forming the points of the star would be half those of the central angles, their measures are $\frac{72°}{2} = 36°$.

Check for Understanding: Investigation 2

1. C
2. B, D
3. Together, the four triangles contain $4 \cdot 180° = 720°$. Since the sum of the measures of the four central angles is 360°, the sum of the measures of the triangles' base angles must be $720° - 360° = 360°$. Suppose the base angles of $\triangle WOX$ measure $a°$, those of $\triangle XOY$ measure $b°$, those of $\triangle YOZ$ measure $c°$, and those of $\triangle WOZ$ measure $d°$. This means that $2a + 2b + 2c + 2d = 360$, or $a + b + c + d = 180$. Thus, since the opposite angles in quadrilateral $WXYZ$ have measures that sum to $a° + b° + c° + d°$, the angles are supplementary.

Check for Understanding: Investigation 3

1.

Side Length (cm)	Apothem Length (cm)	Area (cm²)
14	7	196
16	8	256
24	12	576
18	9	324
20	10	400
12	6	144

2. A
3. Dana could have divided a regular octagon into eight isosceles triangles, each with a vertex angle measuring $\frac{360°}{8} = 45°$. An apothem divides one of these triangles into two right triangles and bisects the vertex angle. Thus, if s is the octagon's side length, then $\tan(22.5°) = \frac{0.5s}{a}$. Solving for s gives $s = 2a \cdot \tan(22.5°)$, so the octagon's perimeter, P, is $P = 8s = 16a \cdot \tan(22.5°)$. Substituting $16a \cdot \tan(22.5°)$ for P in the formula Area $= \frac{1}{2}aP$ and simplifying shows that Dana got the formula Area $= 8a^2 \cdot \tan(22.5°)$, or Area $\approx 3.31a^2$.

Construct and Explore Polygons *(continued)*

Practice Exercises

1. $m\angle A = 70$
   $m\angle B = 90$
2. 16
3. 22.5
4. C
5. $m\angle BPC \approx 51.429$
   $PC \approx 4.637$
   58.835
6. A, C
7. C
8. $x = 70$
   $y = 80$
9. B
10. hexagon

Apply

What Will You Make Your Pasta Look Like?

This problem requires students to make two constructions. One is a convex polygon inscribed in a circle, and the other is a concave polygon with only some of its vertices lying on a circle.

- If you use this problem as a homework assignment, you may want to set up the problem during class time by having students watch the Dies and Extrusions video.
- Students will need a straightedge and compass to complete the constructions.
- As an extension, ask students how much material would be required to produce an extrusion of a certain length.

The Apply problems are rich performance tasks that may have a variety of open-ended solutions. One sample student response for this problem can be found in the Model Lesson.

9.2 Prove and Apply Theorems about Quadrilaterals

Check for Understanding: Investigation 1

1. 360°, 82°
2. C
3. sometimes, always
4. The area of trapezoid *WXYZ* is $\frac{1}{2}h(b_1 + b_2)$ square units. Since $\overline{MN}$ is the trapezoid's midsegment, it is parallel to $\overline{WX}$ and $\overline{ZY}$, and its length is half the sum of their lengths. Also, it divides the trapezoid into trapezoids *WXNM* and *MNYZ*, which have equal heights. This means that the height of trapezoid *MNYZ* is $\frac{1}{2}h$ units, and the lengths of its bases are $\frac{b_1 + b_2}{2}$ and b_2 units. Thus, its area is $\frac{1}{2}\left(\frac{1}{2}h\right)\left(\frac{b_1 + b_2}{2} + b_2\right)$, or $\frac{1}{8}h(b_1 + 3b_2)$ square units.

Check for Understanding: Investigation 2

1. C
2. 8, 16, 3, 3
3. can, cannot
4. Yes, it is possible to know. Since $\overline{ZM}$ bisects $\angle WZY$ and $\overline{YM}$ bisects $\angle XYZ$, $m\angle MZY = \frac{1}{2}m\angle WZY$, and $m\angle MYZ = \frac{1}{2}m\angle XYZ$. Also, consecutive angles of a parallelogram are supplementary, so $m\angle WZY + m\angle XYZ = 180°$. Multiplying each side of the equation by $\frac{1}{2}$ gives $\frac{1}{2}(m\angle WZY + m\angle XYZ) = \frac{1}{2}(180°)$, or $\frac{1}{2}m\angle WZY + \frac{1}{2}m\angle XYZ = 90°$, and substitution gives $m\angle MZY + m\angle MYZ = 90°$. According to the triangle sum theorem, $m\angle MZY + m\angle MYZ + m\angle YMZ = 180°$, so by substitution, $90° + m\angle YMZ = 180°$, and $m\angle YMZ = 180° - 90° = 90°$.

Check for Understanding: Investigation 3

1. C
2. B, D
3. Since quadrilateral *QRST* is a kite, its diagonals, $\overline{QS}$ and $\overline{RT}$, are perpendicular to each other. This means that in $\triangle QRT$, $\overline{QP} \perp \overline{RT}$, and in $\triangle SRT$, $\overline{SP} \perp \overline{RT}$. Suppose that the length of $\overline{QP}$ is x units. In this case, the area of $\triangle QRT$ is $\frac{1}{2}bx$ square units, and the area of $\triangle SRT$ is $\frac{1}{2}b(a - x)$ square units. The area of the kite is the sum of the areas of the two triangles, so it is $\frac{1}{2}bx + \frac{1}{2}b(a - x) = \frac{1}{2}bx + \frac{1}{2}ab - \frac{1}{2}bx = \frac{1}{2}ab$ square units.

Prove and Apply Theorems about Quadrilaterals *(continued)*

Check for Understanding: Investigation 4

1. rhombus, rectangle
2. A, G
3.

Polygon	Parallel Opposite Sides	Congruent Opposite Sides	Congruent Adjacent Sides	Congruent Opposite Angles	Congruent Adjacent Angles
Quadrilateral					
Parallelogram	√	√		√	
Square	√	√	√	√	√
Rhombus	√	√	√	√	
Kite			√	√	
Isosceles Trapezoid	√	√			√
Trapezoid	√				
Rectangle	√	√		√	√

4. Quadrilaterals that could have diagonals that are congruent but not perpendicular are a parallelogram, isosceles trapezoid, trapezoid, and rectangle. Both diagonals of a parallelogram and rectangle, and neither of an isosceles trapezoid or trapezoid, must be bisected, so the minimum for Kurt is zero. Quadrilaterals that could have diagonals that are perpendicular but not congruent are a rhombus and a kite. Both diagonals of a rhombus, and one of a kite, must be bisected, so the minimum for Athena is one.

Check for Understanding: Investigation 5

1. B, D
2. 14, 8
3. The coordinates of W, X, Y, and Z are $(-1, 4)$, $(2, 0)$, $(-1, -4)$, and $(-4, 0)$, respectively. The slope of $\overline{WX}$ is $\frac{0-4}{2-(-1)} = -\frac{4}{3}$, that of $\overline{XY}$ is $\frac{-4-0}{-1-2} = \frac{4}{3}$, that of $\overline{YZ}$ is $\frac{0-(-4)}{-4-(-1)} = -\frac{4}{3}$, and that of $\overline{WZ}$ is $\frac{0-4}{-4-(-1)} = \frac{4}{3}$. The length of $\overline{WX}$ is $\sqrt{(2-(-1))^2 + (0-4)^2} = 5$, that of $\overline{XY}$ is $\sqrt{(-1-2)^2 + (-4-0)^2} = 5$, that of $\overline{YZ}$ is $\sqrt{(-4-(-1))^2 + (0-(-4))^2} = 5$, and that of $\overline{WZ}$ is $\sqrt{(-4-(-1))^2 + (0-4)^2} = 5$. Since side lengths are equal and slopes of adjacent sides are not opposite reciprocals, the most precise term for quadrilateral $WXYZ$ is a rhombus.

Prove and Apply Theorems about Quadrilaterals *(continued)*

Practice Exercises

1. B
2. B
3. Perimeter = 25
   $x = 106.3$
4.

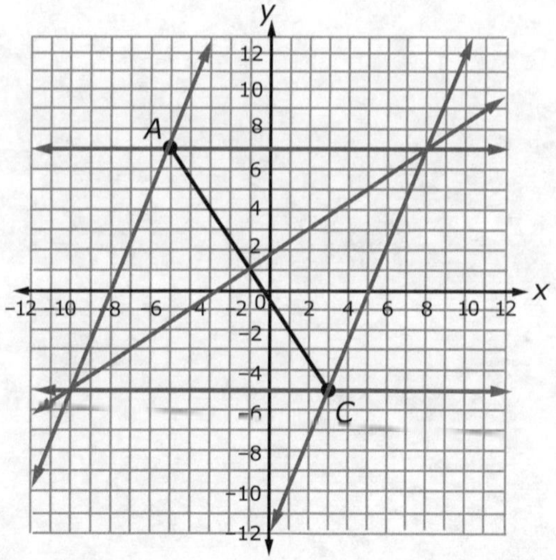

5. B
6. 56.6, 180
7. A
8. B
9. $JL = 12.0416$
   $KM = 12.0416$
   yes
   yes
10. C

Apply

How Would You Organize a County Fair?

This problem offers students opportunities to solve real-world problems by determining lengths of materials needed to build booths for a county fair. Students are challenged to minimize the amount of necessary materials and extend their calculations to determine the amount of materials needed for any number of booths. To minimize the number of different-sized struts used, the diagonals should be congruent.

- If you use this problem as a homework assignment, you may want to set up the problem during class time by having students watch the Surveying a Fairground video.
- You may wish to provide isometric grid paper to help students with their drawings.

The Apply problems are rich performance tasks that may have a variety of open-ended solutions. One sample student response for this problem can be found in the Model Lesson.

10.1 Investigate Cross-Sections and Rotations

Check for Understanding: Investigation 1

1. D, F
2. B
3. could not, could
4. Since the radius of the first cross section is the same as that of the sphere, the plane that created it must have passed through the sphere's center. The length of a line segment from the center of the sphere to any point on the outer edge of the second cross section is 37 cm, since this is the sphere's radius. Suppose the distance between the two planes was d cm. Since the second cross section's radius is 35 cm, $d^2 + 35^2 = 37^2$ by the Pythagorean theorem. Solving for d shows that the two planes were 12 cm apart.

Check for Understanding: Investigation 2

1. D
2. 490π, 700π, 210π
3. In $\triangle RST$, the slope of $\overline{RS}$ is $\frac{0-3}{0-(-3)} = \frac{-3}{3} = -1$, and the slope of $\overline{ST}$ is $\frac{4-0}{4-0} = \frac{4}{4} = 1$. Since the slopes of the sides are opposite reciprocals, $\triangle RST$ is a right triangle. For this reason, it is possible for Mayumi to rotate it about a line to produce a cone. The line about which it is rotated should contain either $\overline{RS}$ or $\overline{ST}$. In other words, the line should have a slope of either -1 or 1, and it should pass through the origin. This means Mayumi can produce a cone by rotating $\triangle RST$ about the line $y = -x$ or about the line $y = x$.

Practice Exercises

1. A, D
2. A, D, E
3. 256
4. A
5. C
6. A
7. B, F
8. C
9. A, C, E, F, G
10. pyramid: B, C
    candle: A, C
    tent: B, C

Investigate Cross-Sections and Rotations *(continued)*

Apply

What Would a Vase Based on Your Profile Look Like?

In this problem, students will use circular cross-sections to build a model of a Rubin's vase, the famous optical illusion that looks like either a vase or two pictures in profile.

- Students may pair up to take each other's pictures. Make sure the photograph is in direct profile against a plain background. You may want to tape some paper to the wall to have a place for pictures.
- If students print the picture too large, it may take a long time to cut enough cardboard circles. One option is to use cardboard spacers between the cross-sections.
- To measure the thickness of the cardboard more accurately, suggest measuring a stack of 10 pieces and dividing the measurement by 10. Calipers may be used if available.

The Apply problems are rich performance tasks that may have a variety of open-ended solutions. One sample student response for this problem can be found in the Model Lesson.

10.2 Develop and Apply Volume Formulas

Check for Understanding: Investigation 1

1. C
2. 8, 0.002, 35, 224
3. is, is
4. The converse is that if the volumes of two three-dimensional objects are equal, then the objects must have the same height and the same cross-sectional area at every level. It is not true. For example, one rectangular prism could have a length of 3 cm, a width of 2 cm, and a height of 8 cm, while another could have a length of 4 cm, a width of 1 cm, and a height of 12 cm. Both would have a volume of 48 cm^3, but they would have different heights (8 cm and 12 cm) and cross-sectional areas (6 cm^2 and 4 cm^2).

Check for Understanding: Investigation 2

1. D
2.

Layer	Base Side Length (in.)	Height (in.)	Volume (in.^{3})
Top	2	1	4
Middle	4	1	16
Bottom	6	1	36

56, 36

3. The formula for the exact volume, V, of a square pyramid is $V = \frac{1}{3}b^2h$, where b is the base side length, and h is the height. Thus, the exact volume of Willard's pyramid is $V = \frac{1}{3}(5^2)(4) = 33\frac{1}{3}$ cm^3. If the approximate volume of 44 cm^3 decreased by 29.6%, it would become $44(1 - 0.296) = 30.976$ cm^3. An approximate volume of 30.976 cm^3 is impossible, because the formula $V \approx b^2h\left(\frac{1}{3} + \frac{1}{2n} + \frac{1}{6n^2}\right)$ will always give an approximate volume greater than the actual volume. Willard must be incorrect in his reasoning.

Check for Understanding: Investigation 3

1. B
2. 18
3. The areas of the cross-sections of the cylinder, cone, and hemisphere are πr^2, πa^2, and πb^2 square units, respectively. This means the area of the cross-sectional ring on the left is $\pi r^2 - \pi a^2$ square units. Using the Pythagorean theorem in the hemisphere gives $a^2 + b^2 = r^2$, or $b^2 = r^2 - a^2$. Multiplying by π gives $\pi b^2 = \pi r^2 - \pi a^2$, so the area of the hemisphere's cross-section equals that of the cross-sectional ring. By Cavalieri's Principle, the hemisphere's volume equals the cylinder's volume minus the cone's volume.

Develop and Apply Volume Formulas *(continued)*

Check for Understanding: Investigation 4

1. D
2. 1,500
3. A, E
4. With the cylinder's diameter at 8 mm, the total volume of the capsules produced was about $75{,}000 \cdot 1{,}022 = 76{,}650{,}000 \text{ mm}^3$. If the cylinder's diameter were 6 mm, then the volume of a capsule would have been about $(\pi)\left(\frac{6}{2}\right)^2(15) + \left(\frac{4}{3}\right)(\pi)\left(\frac{6}{2}\right)^3 \approx 537 \text{ mm}^3$. Thus, the same amount of medicine would have filled about $\frac{76{,}650{,}000}{537} \approx 142{,}737$ capsules, so the company could have produced about $142{,}737 - 75{,}000 = 67{,}737$ more capsules.

Check for Understanding: Investigation 5

1. will not, will not
2. 49π, 175π, 224π
3. B
4. The length of the base of the net's rectangular portion is the circumference of each circular portion of the net, or $2\pi r$ units. Suppose the height of the net's rectangular portion is h units. Since the surface area of a sphere of radius r units is $4\pi r^2$ square units, it must be true that $2\pi rh = 4\pi r^2$. Dividing both sides of the equation by $2\pi r$ gives $h = 2r$, so the height of the net's rectangular portion is $2r$ units. Therefore, the ratio of the length of the base of the net's rectangular portion to its height is $2\pi r : 2r$, or $\pi : 1$.

Practice Exercises

1. B
2. 5,333
   573
   5,906
3. A, B
4. C
5. A, D
6. 33.5, 100.5, 50.3
7. $\frac{1}{3}\pi r^2 h$, 5
8. B
9. 28, 14
10. 0.471

Develop and Apply Volume Formulas *(continued)*

Apply

How Much Would It Cost to "Print" a 3-D Game Piece?

In this problem, students will design a game piece to print on a 3-D printer. They will estimate its volume and determine the amount of filament used to print it, then estimate the cost of printing. They will create a hollow version of the piece and compare the cost of printing to that of the solid piece.

- Note that it is possible to print a solid with a completely enclosed hollow section. However, it may be more cost-effective to leave an opening at the bottom. If possible, have some pieces available from different games to give students ideas for their designs.
- Students may practice designing their pieces using software for 3-D printers. If a 3-D printer is available, have students print the piece.
- As an extension, students working in groups can design a game with multiple pieces and play the game.

The Apply problems are rich performance tasks that may have a variety of open-ended solutions. One sample student response for this problem can be found in the Model Lesson.

11.1 Analyze Graphs and Equations of Circles and Ellipses

Check for Understanding: Investigation 1

1. A

2. $(-2, 1), 3, (x + 2)^2 + (y - 1)^2 = 9$

3. B, C, E

4. You could draw an infinite number of right triangles. The hypotenuse of the right triangle represents the circle's radius, and a circle has an infinite number of radii. In other words, there are an infinite number of points on the circle, so you could draw an infinite number of line segments from the circle's center to a point on the circle.

Check for Understanding: Investigation 2

1. A. VI.
B. V.
C. VIII.
D. VII.
E. I.
F. II.
G. III.
H. IV.

2. $(x + 8)^2 + (y - 5)^2 = 121, (-8, 5), 11$

3. In graphing form, $x^2 + y^2 - 14x + 48y = 1$ is $(x - 7)^2 + (y + 24)^2 = 626$, while $x^2 + y^2 + 30x - 40y = 1$ is $(x + 15)^2 + (y - 20)^2 = 626$, so Brad is correct that the two original equations represent circles with radii of the same length. However, this is just a coincidence. Usually, if two equations in the form $x^2 + y^2 + ax + by = c$ have different values of a and b, but the same value of c, the circles represented by the equations will have different radii lengths.

Check for Understanding: Investigation 3

1. C

2. $a = \sqrt{b^2 - c^2}, b = \sqrt{a^2 + c^2}$ or $\sqrt{c^2 + a^2}$

3. B, C, F

4. Since the ellipse is horizontal and not a circle, the value of c must be greater than 0. Also, the ellipse's foci are closer to the center of the ellipse than the ellipse's vertices, so the value of c must be less than the value of a. Therefore, $0 < \frac{c}{a} < 1$. As the value of $\frac{c}{a}$ gets smaller and smaller, it approaches 0, so the ellipse gets closer and closer to being a circle.

Analyze Graphs and Equations of Circles and Ellipses *(continued)*

Check for Understanding: Investigation 4

1. (2, –6), 7, 3, $\frac{(x-2)^2}{49} + \frac{(y+6)^2}{9} = 1$
2. C
3. The center of the ellipse is given as (−1, 8). Since the distance from the ellipse's center to its vertex is $-1-(-9) = 8$ units, the value of a^2 is $8^2 = 64$, so the ellipse's equation is $\frac{(x+1)^2}{64} + \frac{(y-8)^2}{b^2} = 1$. To find b, substitute the point (−5, 14) into the equation, which results in $\frac{16}{64} + \frac{36}{b^2} = 1$. After eliminating the fractions, the equation is $16b^2 + 2{,}304 = 64b^2$. Solving for b^2 gives $b^2 = 48$, so the equation of the ellipse is $\frac{(x+1)^2}{64} + \frac{(y-8)^2}{48} = 1$.

Check for Understanding: Investigation 5

1. A. IV.
   B. VI.
   C. V.
   D. II.
   E. I.
   F. III.
2. $(1, -5-\sqrt{55}), (1, -5+\sqrt{55})$
3. The equation $3x^2 + 11y^2 - 24x = 18$ should be converted into graphing form. This can be done by separating the x-terms from the y-terms, factoring to get $3(x^2 - 8x) + 11y^2 = 18$, and completing the square to get $3(x-4)^2 + 11y^2 = 66$. Then you can divide by 66 to get $\frac{(x-4)^2}{22} + \frac{y^2}{6} = 1$. Since the foci are $\sqrt{22-6} = \sqrt{16} = 4$ feet from the center, the ball should be placed $4 \cdot 2 = 8$ feet from the hole.

Analyze Graphs and Equations of Circles and Ellipses *(continued)*

Practice Exercises

1. C
2. Center point: (1, 5)
   Longest horizontal line segment: 8
   Longest vertical line segment: 10
3. Center point: (8, 8)
   Length of the longest horizontal line segment: 12
   Length of the longest vertical line segment: $2\sqrt{27}$
4. 4, –6, 0
5. Centered at *A*, passing through *C*: B
   Centered at *B*, passing through *D*: A
   Centered at *C*, passing through *A*: D
   Centered at *D*, passing through *B*: C
6. $x_0 = 5$
   $y_0 = -3$
   $r = \sqrt{18}$
7. $x_0 = -5$
   $y_0 = 1$
   $r = 3$
8. Center: (1, –2)
   Radius: $\sqrt{3}$
9. D, F
10. 3, –2, $\sqrt{149}$
11. A, D, F, H

Apply

Where Am I?

This problem challenges students to use circles to triangulate a specific location. Students use what they know about circles to provide the appropriate equations and use algebra and graphs to prove that their triangulation works.

- Students may upload the graph in the Dynamic Geometry Tool to draw the circles, or they may plot the cities on a sheet of graph paper and use a compass. Students could also choose their own three cities on their own map.
- As an extension, students can exchange and solve each other's problems.

The Apply problems are rich performance tasks that may have a variety of open-ended solutions. One sample student response for this problem can be found in the Model Lesson.

11.2 Analyze Graphs and Equations of Hyperbolas

Check for Understanding: Investigation 1

1. B, E, F
2.

	Number of Units From Point to Focus 1	Number of Units From Point to Focus 2
Point 1	6	14
Point 2	13	21
Point 3	10	18
Point 4	17	25
Point 5	8	16
Point 6	3	11

3. Suppose we call the two vertices of the hyperbola vertex 1 and vertex 2, and the two foci of the hyperbola focus 1 and focus 2. From the given information, we can say that the distance between vertex 1 and focus 1 is 2 units, and the distance between vertex 1 and focus 2 is 9 units. Since the distance between vertex 2 and focus 2 must also be 2 units, the distance between vertex 1 and vertex 2 must be $9 - 2 = 7$ units.

Check for Understanding: Investigation 2

1. D
2. A, E
3. (0, 0), (−5 or 5, 0), (5 or −5, 0), 10, 20, $-2x$ or $2x$, $2x$ or $-2x$
4. If you know the vertices of the auxiliary rectangle, then you can also determine the value of a, the value of b, the coordinates of the center, and the equations of the asymptotes. The only additional information you would need to know is the orientation of the hyperbola. You can then determine the length of the transverse and conjugate axes, as well as the coordinates of the vertices and foci, and you will be able to graph the hyperbola.

Check for Understanding: Investigation 3

1. D
2. $y = \pm\frac{\boxed{12}}{\boxed{5}}(x - \boxed{1}) - \boxed{4}$
3. When the asymptotes are graphed, we can see that they intersect at (−4, 0), so this is the center. Since the vertex is 3 units above the center, the value of b is 3. Also, the vertex is the midpoint of a side of the auxiliary rectangle, so this side must have endpoints at (−8, 3) and (0, 3), meaning that the value of a is 4. Therefore, the hyperbola's equation is $\frac{y^2}{9} - \frac{(x+4)^2}{16} = 1$, and solving for the positive values of y gives $y = 3\sqrt{\frac{(x+4)^2}{16} + 1}$.

Analyze Graphs and Equations of Hyperbolas *(continued)*

Check for Understanding: Investigation 4

1. C
2. (–1, 3), 16, 12, (–1, –7 or 13), (–1, 13 or –7)
3. First, Warren factored –1 from only $-y^2$. He should have factored it out from both $-y^2$ and $8y$. Then, instead of adding 4 and 16 to the right side when completing the squares, he should have added 8 and subtracted 16, because all terms within the first set of parentheses are being multiplied by 2, and all those within the second set are being multiplied by –1. The correct graphing form is $\frac{(x+2)^2}{8} - \frac{(y-4)^2}{16} = 1$.

Practice Exercises

1. D
2. C
3. $a = 5$
   $b = \sqrt{11}$
4. $a = \sqrt{525}$
   $b = 10$
5. $a = 4$
   $b = 3$
   $h = -3$
   $k = -4$
6. $a = 20$
   $b = 15$
   $h = 10$
   $k = -5$
7. C
8. C
9. 5
10. B, D, E
11. 63

Apply

How Are Hyperbolas Used in Designing Structures?

This problem presents students with a real-world example of a hyperboloid, and they find the equation and key features of the 2-D projection, a hyperbola. Students use what they know about hyperbolas to analyze the usefulness of the structure.

- To avoid research, you may wish to provide students with the PDF containing the diagram of the tower.
- If needed, inform students that the tower stands 30 feet tall.
- If you use this problem as a homework assignment, you may want to set up the problem during class time by having students watch the video Apply Problem: Hyperboloid Structures.

The Apply problems are rich performance tasks that may have a variety of open-ended solutions. One sample student response for this problem can be found in the Model Lesson.

11.3 Analyze Graphs and Equations of Parabolas

Check for Understanding: Investigation 1

1. B
2. A, C, F
3. 2, 5
4. If the directrix is the x-axis, the parabola is either entirely above the x-axis and opens up, or entirely below the x-axis and opens down, so it could lie in the first and second quadrants, or in the third and fourth quadrants. If the directrix is the y-axis, the parabola is either entirely left of the y-axis and opens left, or entirely right of the y-axis and opens right, so it could lie in the second and third quadrants, or in the first and fourth quadrants.

Check for Understanding: Investigation 2

1. $\sqrt{(x-8)^2+(y-6)^2}$ or $\sqrt{(y-6)^2+(x-8)^2}$, $y-4$
2. A
3. As shown in the graph, the parabola opens upward, and its vertex is $(-4, -8)$, so its equation must be in the form $y = a(x+4)^2 - 8$, where a is a positive constant. The parabola passes through the point (8, 4). Substituting this point into the equation and solving for a gives $a = \frac{1}{12}$, so the parabola's equation is $y = \frac{1}{12}(x+4)^2 - 8$. Since $a = \frac{1}{4p}$ and $p = 3$, the parabola's focus must be $(-4, -5)$, and its directrix must be $y = -11$.

Check for Understanding: Investigation 3

1. B
2. (–4, 7), (3, 7), $x = -11$
3. C, E
4. The parabola's vertex is half the perpendicular distance from the focus to the directrix, so it is at (12, 1.25). This means $p = 0.25$, and since the directrix is above the focus, the parabola opens down, so the parabola's equation in vertex form is $y = -(x-12)^2 + 1.25$. Rewritten in standard form, the equation is $y = -x^2 + 24x - 142.75$, so $a = -1$, $b = 24$, and $c = -142.75$.

Analyze Graphs and Equations of Parabolas *(continued)*

Practice Exercises

1. −3
2. −17
3. C
4. −6.4
5. −3
6. $-\frac{73}{24}$
7. 4, 5
8. 4
9. A, D, F
10. C
11.

Parabola Opens Up	Parabola Opens Down	Not a Parabola
A	C	E
B	D	F

Apply

Can a Parabola Help You Hear Better?

This problem challenges students to find some type of audio equipment that uses parabolas to enhance sound and use their knowledge of parabolas to diagram the parabola in the equipment's cross section and derive its equation using the focus and directrix.

- If research is not an option, provide students with information about parabolic microphones. They work by concentrating the incoming sound waves at a single point. The sound waves come from a distance and bounce off the parabolic reflector. Because of the parabolic shape, the incoming parallel sound waves bounce back to a single point—the focus of the parabola—making the sound stronger and clearer. A microphone placed at the focus picks up the enhanced sound. One type of parabolic microphone has a 22.5-inch diameter and is 8 inches deep.
- Students may use the Graphing Calculator to graph the cross-section parabola of the sound enhancer.

The Apply problems are rich performance tasks that may have a variety of open-ended solutions. One sample student response for this problem can be found in the Model Lesson.

11.4 Compare Conic Equations

Check for Understanding: Investigation 1

1. B, D, F
2.

Equation	Circle	Ellipse	Hyperbola	Parabola
$3x^2 - 9x + 1 = 12y^2 + 2y + 15$			√	
$2y^2 + y = 10x^2 + 2y^2 - 9x + 26$				√
$5x^2 + 6x - 7 = 18 - y - 5y^2$	√			
$y^2 - 22y - 3 = 42 + 4x - 8x^2$		√		
$4x^2 - 11x - 33 = 6y^2 + 4x^2 - 3x$				√
$3y^2 - 20y + 2x = 17 - 3x^2 + 9y$	√			
$2x^2 + 7x - 50 = 5 - 13y^2 + 2y$		√		
$y^2 + 8y - 36 = 5x^2 + x + 4$			√	

3. C
4. At the beginning of his flowchart, Julius probably did not include a box that said, "Put the x^2 and y^2 terms on the same side of the equation." For example, suppose he was given the equation $5x^2 + y^2 + 2x - y = 3x^2 + y^2 + 1$. If he puts the x^2 and y^2 terms on the same side, the y^2 terms drop out, and he gets $2x^2 + 2x - y = 1$. Since this equation does not have both an x^2 and a y^2 term, it must represent a parabola, just like Julius got in his graph.

Check for Understanding: Investigation 2

1. D
2. B, C
3. The equation of an ellipse with a center at the origin has the form $\frac{x^2}{a^2} + \frac{y^2}{b^2} = 1$, where the major axis has a length of $2a$ units, and the minor axis has a length of $2b$ units. For the ellipse that Julio drew, the value of a is $\sqrt{625} = 25$, so the length of the ellipse's major axis is $25 \cdot 2 = 50$ units. Since Julio represented the sphere to scale, the length of its radius is equal to half the length of the ellipse's major axis, which is $\frac{50}{2} = 25$ units.

Compare Conic Equations *(continued)*

Check for Understanding: Investigation 3

1.

System of Equations	Maximum Number of Solutions
ellipse and ellipse	4
hyperbola and hyperbola	4
parabola and circle	4
parabola and hyperbola	4

2. A, C, E
3. B, D
4. C
5. Since the system includes the equation of a circle and has four solutions, the other equation could be that of a hyperbola. If the hyperbola's equation were in the form $y^2 - 4x^2 = k$, where k is a constant, then solving for y^2 gives $y^2 = 4x^2 + k$, and substituting into the equation of the circle gives $x^2 + 4x^2 + k = 169$. Substituting either -5 or 5 for x and solving for k gives $k = 44$, so the other equation in the system could be $y^2 - 4x^2 = 44$.

Practice Exercises

1. D, F
2. B, C, F
3. C
4. Point 1: (−3.328, 4.992)
   Point 2: (3.328, 4.992)
5. D
6. Point 1: (−6.915, 5.120)
   Point 2: (6.915, 5.120)
7. Intersection point: (5.164, 3.055)
   Value of A: 0.115
8. B
   A
   D
   C
   E
9. D
10. The system will have exactly 4 solutions if $N = 17$.
    The system will have exactly 2 solutions if $N = 16$.
    The system will not have any solutions if $N = 15$.

Apply

What Path Must the Ball Travel to Score?

This problem presents students with the opportunity to use what they have learned about conic sections to identify and research the parabolic path of a projectile and find and interpret the quadratic equation that models the path. If research is not an option, have students use the introductory photo and make measurements and estimates, or provide students with data for punting a football: angle of inclination: 45°; acceleration due to gravity: −32 ft/s^2; initial height: 2 ft; initial velocity: 88 ft/s. As needed, provide students with background information for determining the tangent and cosine of the angle. Students should not confuse the function with a trigonometric function.

The Apply problems are rich performance tasks that may have a variety of open-ended solutions. One sample student response for this problem can be found in the Model Lesson.